NATIONAL 5

CHEMISTRY

WITH ANSWERS

SECOND EDITION

Stephen Jeffrey, Fran Macdonald, John Anderson, Barry McBride & Paul McCranor

HODDER
GIBSON
AN HACHETTE UK COMPANY

The Publishers would like to thank the following for permission to reproduce copyright material:

Photo credits

p.1 (background) and Section 1 running head image © Zhu difeng/stock.adobe.com (inset L) © 2010 Carlos E. Santa Maria All Rights Reserved (inset C) © Georgios Kollidas/stock.adobe.com; **p.2** (left) © 2010 Carlos E. Santa Maria All Rights Reserved (right) © Pixmax/stock.adobe.com; **p.11** (left) © Graham J Hills/SPL (right) © Georgios Kollidas/stock.adobe.com; **p.12** (far left) © Interfoto/Alamy (second left) © Georgios Kollidas/stock.adobe.com (centre) © Archive Pics/Alamy (second right) © World History Archive/Alamy (far right) © Everett Collection Historical/Alamy; **p.35** (top) © Mark Boulton/Alamy (bottom) © Mark Boulton/Alamy; **p.38** (left) © Pam Tait (right) © Pam Tait; **p.47** (background) and Section 2 running head image © sss78/stock.adobe.com (inset L) © Artem Gorohov - Fotolia (inset C) © Anna Lurye - Fotolia (inset R) © Paulista - stock.adobe.com; **p.48** (left) © Alex Segre/Alamy (right) © DigitalGenetics/stock.adobe.com; **p.50** (top) © Pam Tait (bottom) © Pam Tait; **p.51** (top left) © Artem Gorohov - Fotolia (bottom left) © Fotolia (right) Anna Lurye - Fotolia; **p.52** © PAU; **p.57** (top) © Martin Shields/Alamy (bottom) © Pam Tait; **p.58** (left) © James Holmes/Zedcor/SPL (right) © Paulista/stock.adobe.com; **p.61** (top) © xfotostudio/stock.adobe.com (bottom) © chagin/stock.adobe.com; **p.62** © John Anderson; **p.63** (left) © John Anderson (right) © John Anderson; **p.64** (top left) © inga spence/Alamy (top right) © Tony Watson/Alamy (bottom right) © John Anderson; **p.67** © AlenKadr/stock.adobe.com; **p.68** © inga spence/Alamy; **p.71** (top left) © Artem Gorohov - Fotolia (bottom left) © Jason Lindsey/Alamy (right) © Hans Kwaspen - Fotolia; **p.72** (top) Martin Lee/Alamy (bottom) © Pam Tait; **p.73** (left) © Hodder Gibson (right) © Phil Degginger/Alamy; **p.75** © Pam Tait; **p.81** (background) and Section 3 running head image © Eisenhans/stock.adobe.com (inset L) © Scanrail - stock.adobe.com (inset C) © Aleš Nowák - stock.adobe.com (inset R) © valdezrl - stock.adobe.com; **p.82** (top left) © Nikolai Sorokin/adobe.stock.com (top right) © Kayros Studio / stock.adobe.com (bottom left) © adamgilchrist/stock.adobe.com (bottom right) © adisa/adobe.stock.com; **p.87** © Qrt/Alamy; **p.88** © World History Archive/Alamy; **p.92** © Antonio Gravante/adobe.stock.com; **p.96** (top) © Scanrail/stock.adobe.com (bottom) Neil McAllister/Alamy; **p.97** © Alain Vermeulen/stock.adobe.com; **p.104** (left) © Aleš Nowák/stock.adobe.com (right) © DaiPhoto/stock.adobe.com; **p.106** © Yasni/Shutterstock.com; **p.108** © valdezrl/stock.adobe.com; **p.110** (top) © Nigel Cattlin/Alamy (right) © Moris Kushelevitch/Alamy; **p.114** (left) © Andy Buchanan/Alamy (right) © Photos 12/Alamy; **p.115** © jojje11/stock.adobe.com; **p.117** (left) © TopFoto (right) © UPP/TopFoto; **p.123** (top) © Archive Pics/Alamy (bottom) © NASA; **p.124** (left) © Empato/Alamy (right) © Paul Glendell/Alamy; **p.127** (c) Pam Tait; **p.130** (c) Pam Tait.

Cartoons by Emma Golley: **p.17** Figure 3.3 (tug-of-war); **p.39** Figure 5.6; **p.64** Figure 7.13

Acknowledgements

Questions marked with an asterisk (*) are copyright Scottish Qualifications Authority and are reproduced with kind permission.

Every effort has been made to trace all copyright holders, but if any have been inadvertently overlooked the Publishers will be pleased to make the necessary arrangements at the first opportunity.

Whilst every effort has been made to check the instructions of practical work in this book, it is still the duty and legal obligation of schools to carry out their own risk assessments.

Although every effort has been made to ensure that website addresses are correct at time of going to press, Hodder Gibson cannot be held responsible for the content of any website mentioned in this book. It is sometimes possible to find a relocated web page by typing in the address of the home page for a website in the URL window of your browser.

Hachette UK's policy is to use papers that are natural, renewable and recyclable products and made from wood grown in well-managed forests and other controlled sources. The logging and manufacturing processes are expected to conform to the environmental regulations of the country of origin.

Orders: please contact Hachette UK Distribution, Hely Hutchinson Centre, Milton Road, Didcot, Oxfordshire, OX11 7HH. Telephone: +44 (0)1235 827827. Email education@hachette.co.uk. Lines are open from 9 a.m. to 5 p.m., Monday to Friday. You can also order through our website: www.hoddereducation.co.uk. If you have queries or questions that aren't about an order, you can contact us at hoddergibson@hodder.co.uk

First published in 2013 © Stephen Jeffrey, Fran Macdonald, John Anderson, Barry McBride, Paul McCranor
This second edition published in 2018 by
Hodder Gibson, an imprint of Hodder Education,
An Hachette UK Company
211 St Vincent Street
Glasgow G2 5QY

Without Answers
Impression number	5	4	3	2	1
Year	2022	2021	2020	2019	2018

ISBN: 978 1 5104 2926 0

With Answers
Impression number	5	4
Year	2022	2021

ISBN: 978 1 5104 2919 2

SCOTLAND EXCEL
We are an approved supplier on the Scotland Excel framework.
Schools can find us on their procurement system as:
Hodder & Stoughton Limited t/a Hodder Gibson.

Cover photo: © Shutterstock / SOMMAI

Typeset in Minion Regular 11/14 by Integra Software Services Pvt. Ltd., Pondicherry, India

Printed in India

A catalogue record for this title is available from the British Library

Contents

Preface

This book is designed to act as a valuable resource for students studying National 5 Chemistry. It provides a core text that adheres closely to the SQA course specification. Each section of the book matches an area of the specification; each chapter addresses a key area of skills or knowledge and understanding of the course. In addition to the core text, the book contains a variety of special features: *Activities*, *Questions* and *Checklists for Revision*. Questions marked with an asterisk (*) are from past SQA exam papers.

SQA assessment for National 5

Course assessment and the assignment

At the end of the National 5 Chemistry course, you will sit a written test. The question paper will be worth a total of 100 marks and it will have two sections:

- Section 1 will contain 25 objective test (multiple choice) items and be worth 25 marks.
- Section 2 will contain restricted and extended response questions and be worth 75 marks.

Open-ended questions

Section 2 will contain two open-ended questions, each of which will be worth 3 marks. Open-ended questions can be used to assess whether or not you have truly grasped a chemical concept. In this type of question, you are required to draw on your understanding of key chemical principles to solve a problem or challenge. The open-ended nature of these questions is such that there is no unique correct answer. In addition to testing the extent of your chemical insight, these questions promote and reward creativity and analytical thinking. Your response to an open-ended question will be assigned a mark according to how well it conveys the depth of your understanding of chemistry. More information on open-ended questions will be found in the Appendix on page 134 of this book.

The assignment

Towards the end of your National 5 course you will be required to carry out an assignment. The purpose of the assignment is to allow you to carry out an in-depth study of a chemistry topic. You will choose the topic and then be allowed:

- a fixed amount of time in which to investigate/research the underlying chemistry, and
- a fixed amount of time in which to write your final report.

The assignment will be worth 20 marks. Full details of the requirements of this task can be found in the SQA 'Instructions for candidates' document which your teacher or lecturer will provide.

Section 1

Chemical Changes and Structure

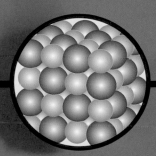

Rates of reaction

Imagine a world without chemistry. Your first thought might be that you would have a free period on your timetable at school, or that you would have picked a different subject instead. That probably is not all that appealing since you have chosen to study chemistry. Maybe you think it is a good thing though, that you would never need to wear safety goggles. Well, you would be right … sort of.

You would never need to wear safety goggles, because you would not be here. You would not exist. Nearly everything around you right now – your clothes, your bag, the products in your hair, the stool you are probably sitting on – would never have existed. The vast majority of the materials in our world were made by chemical reactions.

Chemical reactions are responsible for the world looking and progressing the way it does. Some reactions are over very quickly, like the burning of fuels or explosions (Figure 1.1), while others can take minutes, hours, days, months (Figure 1.2) or even millions of years.

The first part of this section is about how chemists control the **rate** at which reactions take place. You may have learned about this in your lower school science lessons when you found out how to change the speed of a reaction, or in the National 4 Chemistry

Figure 1.2 Rusting is a slower reaction

course when you began to use the word 'rate' instead of 'speed'. In National 5 you will learn how to measure the rate of a reaction and express it as a number with a unit. This is a bit like saying a car is travelling at 70 miles per hour rather than simply that a car is travelling fast.

To change the rate of a reaction, in other words to speed it up or slow it down, it is necessary to change the reaction conditions. There are four **variables** that, when changed, will affect the rate of a chemical reaction:

- temperature
- concentration
- particle size
- use of a catalyst.

For any chemical reaction to take place, the reacting particles must collide with enough energy in order to combine and form new products. Changing one of the four variables above changes the way that these collisions take place, either by causing more collisions to occur, by increasing the energy of each collision, or both.

Temperature

If the temperature of a reaction mixture is increased, the particles have more energy and so move more quickly (Figure 1.3). Increasing the temperature increases the rate of reaction because the particles collide with more energy.

Figure 1.1 Explosions are fast reactions

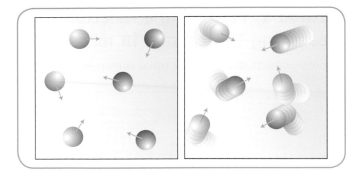

Figure 1.3 Particles at a higher temperature move more quickly so collide with greater energy

In summary:

- The higher the temperature, the faster the reaction.
- The lower the temperature, the slower the reaction.

Concentration

This variable can only be changed for reactants that are in solution. If the concentration of the reactants is increased, there are more reactant particles moving around in the same space (Figure 1.4). There will be more collisions and so the reaction rate is increased.

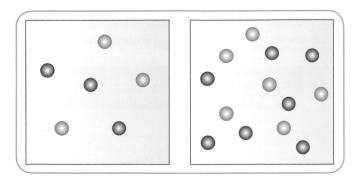

Figure 1.4 At a higher concentration, there are more collisions

In summary:

- The higher the concentration, the faster the reaction.
- The lower the concentration, the slower the reaction.

Particle size

This variable can only be changed for solid reactants. By crushing lumps into powder, for example, we decrease

the particle size of a reactant and increase its surface area. The greater the surface area, the higher the chance of collisions occurring and thus the faster the rate of reaction.

For example, think of a cube where the length of every side is 2 cm (Figure 1.5).

The area of one face of the cube will be $2 \times 2 = 4\,cm^2$.

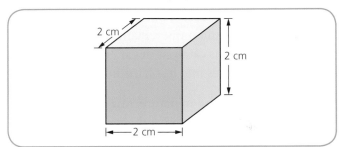

Figure 1.5

The cube has six faces, so the total surface area is $4\,cm^2 \times 6 = 24\,cm^2$.

We could cut that cube horizontally and vertically along each face so that we have eight smaller cubes (Figure 1.6).

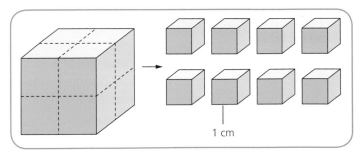

Figure 1.6

Each of the small cubes has a face area of $1\,cm \times 1\,cm = 1\,cm^2$.

The six faces give a total surface area for each smaller cube of $6\,cm^2$.

There are eight cubes so the total surface area is $6\,cm^2 \times 8 = 48\,cm^2$.

In summary:

- The smaller the particle size, the faster the reaction.
- The larger the particle size, the slower the reaction.

Catalysts

Catalysts are used a lot in the chemical industry as they allow reactions to be carried out at lower temperatures, saving both energy and money. Platinum, rhodium and palladium are used as catalysts in car exhaust systems to convert harmful gases into less harmful gases.

Catalysts are also found in nature. **Enzymes** are biological catalysts and some of them are used within the body. For example, amylase in saliva helps to break down starch quickly to produce sugars which can give us energy. Enzymes can also be used in industrial processes such as the **fermentation** of sugars to produce alcohol.

Catalysts are substances that speed up chemical reactions but can be recovered chemically unchanged at the end of the reaction.

Measuring the rate of a reaction

The rate of a chemical reaction is a measure of how fast the reactants are being used up or how fast the products are being made. To measure the rate of any reaction accurately, we can follow one of three changes:

- changes in the mass of the reactants or products
- changes in the volume of the reactants or products
- changes in the concentration of the reactants or products.

Consider the reaction between chalk lumps and hydrochloric acid. This reaction produces carbon dioxide gas, which gives us two opportunities to monitor the course of the reaction and thus measure the rate of reaction.

Method 1: Following a reaction by measuring the change in mass

The experiment should be set up as shown in Figure 1.7.

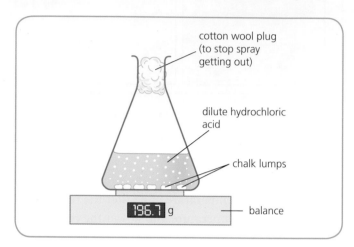

Figure 1.7 Measuring the rate of the reaction between chalk lumps and dilute hydrochloric acid

As the carbon dioxide gas is produced, it leaves the solution and so the total mass of the apparatus goes down. This loss in mass represents the mass of the carbon dioxide gas that has been produced and has escaped. If the initial mass of the apparatus was recorded and new mass readings were taken every 30 seconds, the results shown in Table 1.1 could be obtained.

Time (s)	Mass of apparatus (g)
0	198.1
30	196.7
60	195.8
90	195.4
120	195.0
150	194.7
180	194.4
210	194.1
240	193.9
270	193.8
300	193.8

Table 1.1

These results can be used to plot a graph (Figure 1.8) that shows how the rate of reaction changes with time. The graph can then be used to calculate the actual rate of the reaction. We can calculate the average rate of the reaction from start to finish, or we can calculate an average rate over any time period as the reaction is proceeding.

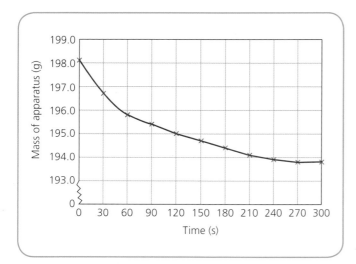

Figure 1.8

Method 2: Following a reaction by measuring the change in volume

The same experiment could be carried out using different apparatus that would allow us to measure the volume of carbon dioxide that is produced (Figures 1.9 and 1.10).

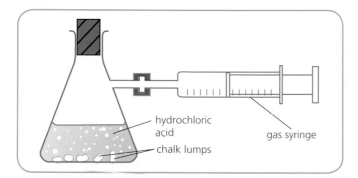

Figure 1.9 Using a gas syringe to monitor the volume of gas produced in a reaction. This method is used if the gas produced is soluble in water

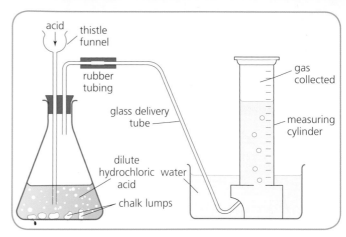

Figure 1.10 Collecting the gas produced in a reaction over water

If the volume of gas produced was noted every 30 seconds, the results in Table 1.2 could be obtained.

Time (s)	Volume of CO_2 produced (cm^3)
0	0
30	28
60	39
90	46
120	51
150	55
180	58
210	60
240	61
270	62
300	62

Table 1.2

A graph of these results would look like Figure 1.11. The section of the graph between 0 and 30 seconds has the highest gradient. The steepness of the line on the graph is an indication of the rate of reaction. This graph shows us that carbon dioxide is being produced very rapidly during the first 30 seconds.

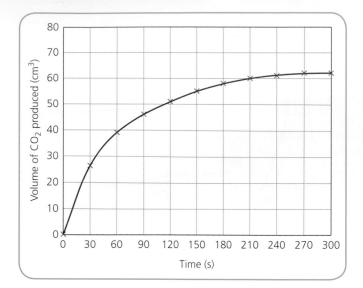

Figure 1.11

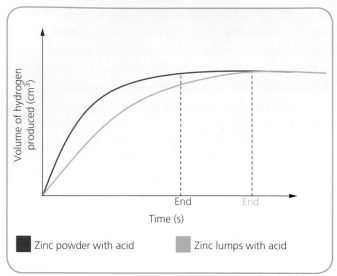

Figure 1.12

Later in the reaction, say between 60 and 180 seconds, we can see that the line is less steep. Carbon dioxide is still being produced, but not as quickly, and so the gradient is less than in the initial section of the graph.

Finally, between 270 and 300 seconds, the graph is horizontal – it has zero gradient. This means that the reaction has stopped and no more carbon dioxide is being produced.

The rate of the reaction is high to begin with because the concentration of the reactants is at its greatest; gradually, the rate slows down as the concentration of the reactants decreases.

Drawing graphs

Instead of an accurately plotted graph that can be used to calculate rate, sometimes you are provided with a sketch of a graph that shows the course of a chemical reaction. While these sketches do not always have numbers and units on their axes, they still provide information about the rate of the reaction as it proceeds.

Figure 1.12 shows two similar reactions between zinc metal and acid. This reaction produces hydrogen gas, the volume of which can be measured over time.

From the graph, we can see that at the start of the reaction, near the origin of the time axis, the red line (which represents zinc powder reacting with dilute

hydrochloric acid) indicates a faster reaction than the green line (the same reaction, but with zinc lumps used instead of powder) as the line is steeper. The end point of both reactions is clearly visible where the lines become horizontal, showing that no further hydrogen gas is being produced. The reaction involving zinc powder, as we would expect, finishes before that with the zinc lumps.

Calculating the average rate of a reaction

Over the course of a chemical reaction, the average rate can be calculated. As previously stated, we can calculate the average rate of the reaction from start to finish or we can calculate an average rate over any time period as the reaction is proceeding.

The average rate of a reaction is always given by the formula:

$$\text{average rate} = \frac{\text{change in measurable quantity}}{\text{change in time}}$$

This can also be written as:

$$\text{average rate} = \frac{\Delta \text{ measurable quantity}}{\Delta \text{ time}}$$

The 'measurable quantity' we use can be the concentration, mass or volume of the reactants or products. As the rate of the reaction is constantly changing throughout the process, this calculation gives us the average rate over a specific period of time.

Worked example

Using the data from Table 1.1, calculate the average rate of the reaction between chalk lumps and acid between 30 and 120 seconds.

Reading from the table, at 30 seconds the mass of the apparatus was 196.7 g and at 120 seconds it was 195.0 g.

$$\text{average rate} = \frac{\text{change in mass}}{\text{change in time}}$$

$$= \frac{(196.7 - 195.0)}{(120 - 30)}$$

$$= \frac{1.7}{90}$$

$$= 0.019\,\text{g s}^{-1}$$

Significant figures, decimal places and rounding

On a basic calculator, dividing 1.7 by 90 will show as 0.01888889. The answer given as 0.019 has been rounded to 2 significant figures or 3 decimal places (dps). An answer of 0.02 (1 significant figure or 2 dps) would also be acceptable.

Be careful when rounding. For example, 1.249 rounded to 1 dp would be 1.2. You cannot round to 2 dps first (giving 1.25), and then further round to 1 dp (giving 1.3). An *initial* figure of 1.25 rounded to 1 dp is 1.3.

Throughout this book, no hard and fast rule has been applied – most numerical answers are given to 1 decimal place, but occasionally, 2, 3 and sometimes even 4 decimal places are used. When writing your final answer in an examination, you should indicate the extent of any rounding in brackets to be on the safe side. Generally, an answer given to 2 dps is acceptable.

It is important to note that the units for a rate calculation change depending on the quantity that was measured. In the above example, the mass was measured in grams and the time in seconds, so the average rate is given the unit grams per second (g s^{-1}). You may also see this written as g/s.

Worked example

The volume of oxygen gas produced when hydrogen peroxide decomposes was measured and the results were plotted to give the graph shown in Figure 1.13.

Calculate the average rate of reaction over the first 25 seconds of the reaction.

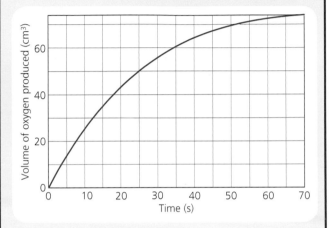

Figure 1.13

$$\text{average rate} = \frac{\text{change in volume}}{\text{change in time}}$$

$$= \frac{(50 - 0)}{(25 - 0)}$$

$$= \frac{50}{25}$$

$$= 2\,\text{cm}^3\,\text{s}^{-1}$$

The unit in this example is cubic centimetres per second (cm^3 s^{-1}) as the volume of oxygen was measured in cubic centimetres and the time in seconds.

Checklist for Revision

I can calculate the average rate of a reaction from graphs or tables of data to show the changes in the rate of a reaction as it progresses.

End-of-chapter questions

1 The course of a chemical reaction was followed by measuring the mass lost by the apparatus over time. The mass decreased by 50 grams in 200 seconds. What was the average rate of reaction in $g\,s^{-1}$?

 A 0.25

 B 0.333

 C 1

 D 4

2 Line A in Figure 1.14 refers to the reaction between magnesium and $0.1\,mol\,l^{-1}$ nitric acid. What change in the reaction conditions could result in the reaction represented by line B?

 A Use magnesium lumps instead of powder.

 B Heat the acid.

 C Add a catalyst.

 D Stir the reaction mixture.

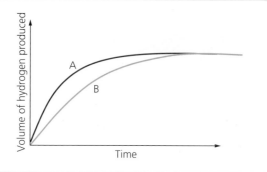

Figure 1.14

3 Hydrogen peroxide breaks down to form oxygen gas and water. You are asked to carry out this reaction and measure the reaction rate. Which combination of pieces of apparatus in Figure 1.15 would be *most* suitable to use in order to calculate the rate of the reaction?

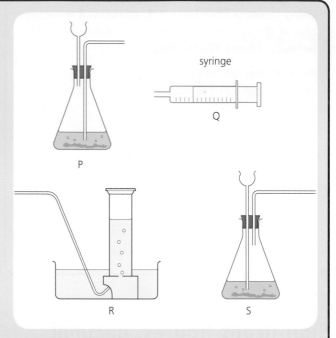

Figure 1.15

 A P and Q

 B S and Q

 C P and R

 D S and R

4 A student was following the rate of the reaction between chalk lumps and dilute acid by measuring the volume of carbon dioxide gas released over time. Which unit should she use when calculating the rate?

 A $cm^3\,s^{-1}$

 B $g\,s^{-1}$

 C $mol\,l^{-1}\,s^{-1}$

 D $°C\,s^{-1}$

5 Figure 1.16 shows how the concentration of a reactant changes over time as products are being formed. What is the average rate of reaction between 20 seconds and 40 seconds in $mol\,l^{-1}s^{-1}$?

A 0.001875

B 0.0025

C 0.00375

D 0.005

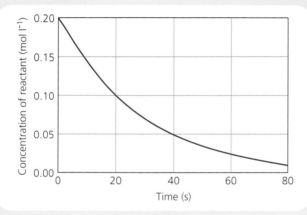

Figure 1.16

6 Margarine is made in the chemical industry by the reaction between vegetable oil and hydrogen gas. To help the reaction to work effectively, the vegetable oil is heated and a powdered nickel catalyst is used. Suggest why the nickel catalyst is used as a powder rather than as a lump.

7 A student was investigating the rate of the reaction between copper(II) carbonate powder and dilute hydrochloric acid. He decided to measure the mass of carbon dioxide gas that was produced over time. He obtained the results given in Table 1.3.

a) Draw a diagram of the apparatus that he could have used to carry out the reaction.

b) Plot a line graph of the results shown above.

c) With reference to the graph, how could he tell that the reaction had finished?

d) Calculate the average rate of the reaction over the first 10 seconds.

Time (s)	Mass of carbon dioxide produced (g)
0	0
10	0.24
20	0.33
30	0.37
40	0.39
50	0.39

Table 1.3

8 Zinc lumps react with dilute hydrochloric acid in a test tube producing bubbles of hydrogen gas.

a) What chemical test could be used to identify that the gas produced was hydrogen?

b) Suggest two changes to the conditions that would result in a slower chemical reaction.

c) Copy the graph in Figure 1.17 and add a line to represent the slower reaction.

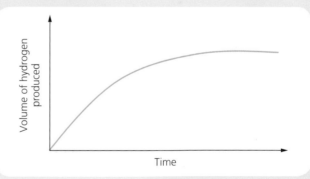

Figure 1.17

9 Figure 1.18 shows the change in the concentration of an acid as a chemical reaction occurs. Use the graph to calculate:

a) the average rate of reaction in $mol\,l^{-1}\,min^{-1}$ over the first 2 minutes

b) the average rate of reaction in $mol\,l^{-1}\,min^{-1}$ between 2 and 7 minutes.

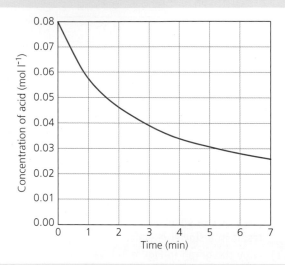

Figure 1.18

Time (s)	Volume of carbon dioxide produced (cm^3)
0	0
20	21
40	35
60	42
80	48
100	52
120	
140	57
160	59
180	60

Table 1.4

10 A student was investigating the rate of reaction between hydrochloric acid and chalk. She weighed out 5 g of chalk lumps, placed them in $0.1\,mol\,l^{-1}$ hydrochloric acid and measured the volume of carbon dioxide produced at regular time intervals. After 180 seconds, the reaction was finished and no chalk remained. Her results are shown in Table 1.4.

a) Plot a line graph of these results.

b) Use your graph to predict the volume of carbon dioxide produced after 120 seconds.

c) Add a dotted line to your graph to show what would happen if 5 g of chalk powder had been used instead of lumps.

d) Draw a labelled diagram of the apparatus the student could have used.

2 Atomic structure

Our world is mostly made up of **compounds** made from different combinations of **elements**. If we looked closely at an element through an extremely powerful microscope, such as a scanning electron microscope, we would see that an element is made of very tiny particles called **atoms**. Figure 2.1 shows gold atoms as seen through a scanning electron microscope.

In this chapter, we will look at atoms and what holds them together.

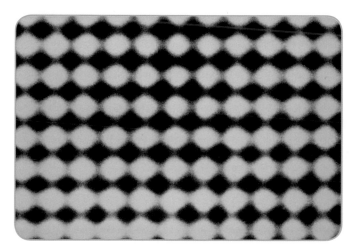

Figure 2.1 Gold atoms as seen through an electron microscope

The atom

Around 460BC Democritus, a Greek scientist, stated that all matter is made of atoms. In 1803 John Dalton (Figure 2.2), an English chemist, presented his atomic theory to the Royal Institution.

Dalton believed that atoms of different elements would have different masses. Based on this, he made some suggestions about atoms:

1 Elements are made from tiny particles called atoms.
2 All the atoms of one element are identical.
3 The atoms of one element are different from the atoms of any other element.
4 Atoms of one element can join with atoms of other elements to form compounds.

In the early twentieth century, however, scientists discovered that atoms are actually made up of even smaller sub-atomic particles called **protons**, **electrons** and **neutrons**.

Several scientists developed theories as to how these particles were arranged in the atom. In 1904 J. J. Thomson, who had discovered the electron in 1897, suggested that an atom was shaped like a sphere with the charged particles (electrons and protons) spread through it like the currants in a plum pudding, but this was proved to be incorrect. Another scientist, Sir Ernest Rutherford, suggested that the positive particles known as protons were clustered in the middle of the atom and he called this collection of particles the **nucleus**. This did not explain, however, why the mass of a single atom of most elements was much greater than the mass of the

Figure 2.2 John Dalton

protons in its nucleus and he suggested that there must be another particle that had yet to be discovered. The neutron was eventually discovered in 1932 by J. Chadwick. Neutrons are also found in the nucleus and they account for the extra mass. In 1913, a Danish scientist called Niels Bohr developed a theory of electron structure which described how the electrons are arranged around the nucleus in shells.

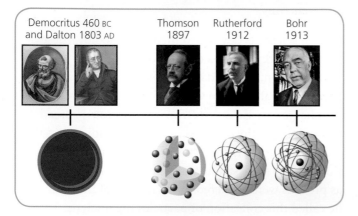

Figure 2.3 Timeline of the history of the atom

The work of these scientists has resulted in the atomic theory that we use today.

Atoms are made up of three smaller particles (Table 2.1):

Electrons: Negatively charged particles that spin around the positive centre of the atom in circles called energy levels. (Imagine how the Moon spins around the Earth.) Their mass is so small it is nearly zero.

Protons: Positively charged particles found in the nucleus of the atom (the centre). They are said to have a relative mass of 1 atomic mass unit.

Neutrons: Neutral particles (they have no electric charge) also found in the nucleus of the atom. They also have a relative mass of 1 atomic mass unit.

Particle	Relative mass	Charge	Location
electron	0	one negative (1−)	energy level
proton	1	one positive (1+)	nucleus
neutron	1	no charge	nucleus

Table 2.1 A summary of sub-atomic particles

The nucleus has an overall positive charge because it contains only protons (positive) and neutrons (neutral) (Figure 2.4).

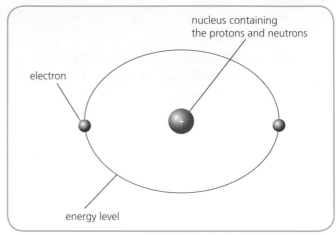

Figure 2.4 The basic structure of an atom

Atoms have no overall charge – they are neutral. This is because they contain equal numbers of positive protons and negative electrons. These opposite charges balance each other making the atom neutral.

The elements are arranged in the Periodic Table in order of increasing **atomic number** – hydrogen has the atomic number 1, helium 2, lithium 3 etc.

The atomic number of an element tells you how many protons there are in an atom of that element.

For example, a sodium atom has 11 protons (Figure 2.5).

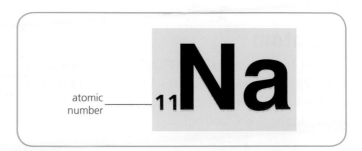

Figure 2.5 Sodium

Since atoms are neutral, this atom must also have 11 negatively charged electrons to balance the 11 positively charged protons. As you learned previously, electrons are found in energy levels around the nucleus of an atom. Each energy level can only hold a certain number of electrons. The first energy level (the one nearest the nucleus) can hold a maximum of two electrons with the others being able to hold up to a maximum of eight electrons. (This is only true for the first 20 elements.)

A sodium atom has 11 electrons, so two go into the first energy level (the one nearest the nucleus) and eight go into the second energy level, which leaves one that must go into the third or outer energy level (Figure 2.6). This information can be summarised as an electron arrangement 2,8,1.

The electron arrangements of other atoms can be found in the Data Booklet.

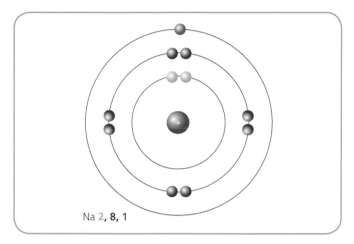

Na 2, 8, 1

Figure 2.6 How the electrons are arranged in a sodium atom

Elements in the same group of the Periodic Table (Figure 2.7) react in similar ways. This is because their atoms all have the same number of outer electrons. All alkali metals, for example, have one outer electron in their atoms (Figure 2.8). This makes them very reactive.

Group 1 – the alkali metals
Li 2,1
Na 2,8,1
K 2,8,8,1
Rb 2,8,18,8,1
Cs 2,8,18,18,8,1
Fr 2,8,18,32,18,8,1

Figure 2.8 Elements in the same group have the same number of outer electrons

Activities

Draw diagrams similar to Figure 2.6 to show the electron arrangement of each of the first 20 elements. Use the SQA Data Booklet to help you.

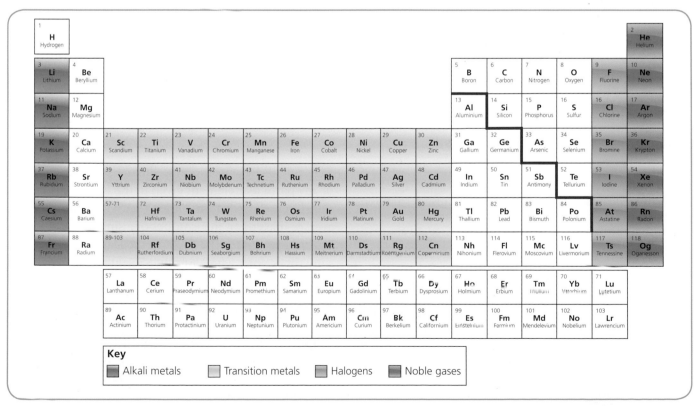

Figure 2.7 The Periodic Table: the red wavy line divides the elements which are metals (to the left of the line) from those which are non-metals (to the right of the line)

Now that we can use the atomic number to state the number of protons and electrons in an atom, we must also be able to work out how many neutrons there are in the nucleus of an atom. To do this we need the **mass number** of the atom.

The mass number of an atom is equal to the number of protons plus the number of neutrons.

The mass number is given at the top left of the element's symbol; for example, the sodium atom in Figure 2.9 has a mass number of 23.

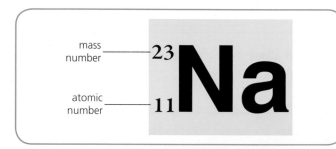

Figure 2.9 Sodium

This way of displaying the mass number, atomic number and symbol for an atom is known as the nuclide notation and it can be used to calculate the numbers of protons, electrons and neutrons present. Figure 2.9 shows that the atomic number of sodium is 11. This tells us that a sodium atom has 11 protons and because its atoms are neutral, it also has 11 electrons.

The mass number of this sodium atom is 23.

Mass of protons = 11

Mass of electrons = 0 (electrons have almost zero mass)

Then the number of neutrons must be 12 to give sodium a mass of 23. (23 – 11 protons = 12 neutrons)

The number of each particle present can be worked out using your P_{rotons} $E_{lectrons}$ $N_{eutrons}$!

$^{19}_{9}F$

P = 9
E = 9
N = 10 (19 – 9)

$^{27}_{13}Al$

P = 13
E = 13
N = 14 (27 – 13)

Some more examples are shown in Table 2.2.

Element	Atomic number	Mass number	Protons	Electrons	Neutrons
magnesium	12	24	12	12	12
potassium	19	39	19	19	20
carbon	6	12	6	6	6

Table 2.2

Isotopes

Think about the boys in your class. They may all be the same age, but they probably all have different masses. This is also true of atoms. Atoms of the same element can have slightly different masses.

Isotopes are atoms with the same atomic number but different mass numbers.

For example: $^{12}_{6}C$ and $^{13}_{6}C$

These two carbon atoms are isotopes; they have the same number of protons but different numbers of neutrons.

For carbon-12:
P = 6
E = 6
N = 6 (12 – 6)

For carbon-13:
P = 6
E = 6
N = 7 (13 – 6)

All elements have isotopes, so the mass of one atom of any particular element given in the Data Booklet is called the **relative atomic mass (RAM)**, which is the average mass of all the isotopes of a single element. The value given for the RAM in the Data Booklet is sometimes not a whole number because it is an average.

The relative atomic mass can be used to give an indication of the relative abundance of an element's isotopes.

Copper has two isotopes: copper-63 and copper-65. The RAM of copper is 63.5. This indicates that copper-63 is the more abundant isotope due to the fact that 63.5 is closer to 63 than to 65.

Bromine has two isotopes: bromine-79 and bromine-81. The RAM of bromine is 80. This indicates that there are equal amounts of both isotopes, because the average of 80 is exactly half way between the masses of the two isotopes.

Ions

Atoms are neutral because they have an equal number of positive protons and negative electrons. When there is an imbalance of electrons to protons, an **ion** is formed. Ions are charged particles and they are formed when atoms lose or gain electrons. Metal atoms lose electrons to form positive ions and non-metal atoms gain electrons to form negative ions (Table 2.3). Why should they do this?

The noble gases are a group of extremely unreactive elements. Apart from helium, which has atoms with only two electrons that fill the first energy level, the electron arrangements for all the noble gases end in 8 – the outer energy level is full. This is said to be a stable electron arrangement and it is the reason why the noble gases are so unreactive. When atoms of other elements combine, their electron arrangements become like those of the noble gases – they achieve a full outer energy level. Neon atoms, for example, have eight outer electrons, whereas oxygen atoms only have six. To become stable like a noble gas, therefore, an oxygen atom must gain two electrons from another atom.

Metals	Non-metals
Magnesium	Chlorine
A magnesium atom has the electron arrangement of 2,8,2. Being a metal, it will lose two electrons to achieve the same electron arrangement as neon, 2,8. This can be shown as an **ion–electron half equation**.	A chlorine atom has the electron arrangement of 2,8,7. Being a non-metal, it will gain one electron to achieve the same electron arrangement as argon, 2,8,8. This can be shown as an ion–electron half equation.
$$Mg \longrightarrow Mg^{2+} + 2e^-$$	$$Cl + e^- \longrightarrow Cl^-$$
This change creates an imbalance in the electron to proton ratio. For the ion $^{24}_{12}Mg^{2+}$:	This change creates an imbalance in the electron to proton ratio. For the ion $^{35}_{17}Cl^-$:
P = 12 E = 10 N = 12	P = 17 E = 18 N = 18

Table 2.3

Checklist for Revision

- I can use nuclide notation for atoms and ions to calculate the numbers of protons, neutrons and electrons.
- I know what isotopes are and that the relative atomic mass (RAM) is the average mass of an atom of an element calculated from data about the element's isotopes and their proportions.

End-of-chapter questions

1 Isotopes of an element have a different number of
   A neutrons
   B protons
   C electrons
   D electrons and protons.

2 An atom of an element has the atomic number of 14 and a mass number of 28. The number of electrons in its atoms is
   A 7
   B 14
   C 28
   D 42

3 Which of the following electron arrangements is that of an element which has similar chemical properties to potassium?
   A 2,8,1
   B 2,8,2
   C 2,8,3
   D 2,8,4

4 Different atoms of the same element have identical
   A mass numbers
   B atomic numbers
   C numbers of neutrons
   D nuclei.

5 Which of the following numbers is the same for both lithium and fluorine atoms?
   A mass number
   B atomic number
   C number of outer electrons
   D number of occupied energy levels

6 The nuclide notation can be used to calculate the number of protons, neutrons and electrons that an ion contains. The nuclide notation of a silver ion is $^{107}_{47}Ag^+$.
   a) Copy and complete Table 2.4 to show the number of each particle that this ion contains.

Particles	Number
protons	
electrons	
neutrons	

   Table 2.4

   b) A sample of silver is found to contain atoms of silver with different masses: ^{107}Ag and ^{109}Ag. What name is given to atoms of the same element with different masses?

   c) The relative atomic mass of silver is 108. What does this suggest about the relative abundance of these different atoms?

7 Table 2.5 shows two isotopes of chlorine.

Isotope	Number of protons	Number of electrons	Number of neutrons
$^{35}_{17}Cl$			
$^{37}_{17}Cl$			

   Table 2.5

   a) Copy and complete the table for each of the isotopes.
   b) Chlorine has the relative atomic mass of 35.5. What does this suggest about the relative abundance of the isotopes?

8 With reference to a Periodic Table, copy and complete Table 2.6.

Isotope	Mass number	Atomic number	Number of protons	Number of electrons	Number of neutrons
$^{31}_{15}P$					
		19			20
	15		7		

   Table 2.6

9 The nuclide notation shows the atomic number and mass number of an isotope. The nuclide notation for an isotope of neon is: $^{21}_{10}Ne$.
   a) An isotope of calcium has the atomic number of 20 and a mass number of 41. Write the nuclide notation for this isotope of calcium.
   b) How many protons and neutrons does this isotope contain?
   c) All the isotopes of calcium are electrically neutral. What does this suggest about the proton to electron ratio of each isotope?

10 Use a Periodic Table to help you copy and complete Table 2.7.

Ion	Number of protons	Number of electrons	Number of neutrons
$^{35}_{17}Cl^-$			
	8	10	8
$^{40}_{20}Ca^{2+}$			

   Table 2.7

3 Bonding, structure and properties

Molecules

We will start this chapter by looking at the chemical bonds that hold atoms together in **molecules**. Molecules are usually made from non-metal atoms only. Water exists as molecules with the chemical formula H_2O. Each molecule contains two hydrogen atoms and one oxygen atom (Figure 3.1). The atoms in a molecule are held in place by **covalent bonds**.

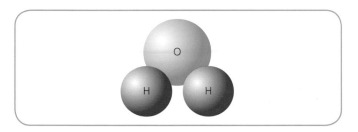

Figure 3.1 A simplified diagram of a water molecule

Covalent bonds

In Chapter 2 we looked at the noble gases and their full outer energy levels of electrons and discussed the idea that atoms of other elements achieve similar electron arrangements when they combine. When non-metal atoms combine with other non-metal atoms, the bonds formed are called covalent bonds.

A covalent bond is a shared pair of electrons between two non-metal atoms.

The atoms are held together because of the electrostatic force of attraction between the positive nuclei of each atom and the negatively charged electrons. This is better illustrated as a diagram (see Figure 3.2).

The saying 'opposites attract' can be used to describe how a covalent bond works. The positively charged nucleus of each atom is attracted to the negatively charged electrons. This creates a 'tug-of-war' effect. Both nuclei try to pull the

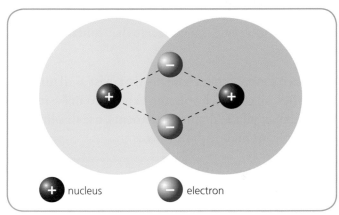

Figure 3.2 Covalent bonding in a molecule

electrons towards themselves, creating a strong bond that holds the atoms together. This is shown by the dashed lines in Figure 3.2.

Figure 3.3 represents a molecule of hydrogen. Hydrogen has the electron arrangement of 1. So if a hydrogen atom shares its electron with that of another atom of hydrogen, they both have two outer electrons, which results in them having the same electron arrangement as the noble gas helium.

Figure 3.3 Covalent bonding can be likened to a tug of war

Remember the tug of war description; it will help you to remember how to describe exactly how a covalent bond holds atoms together.

Hydrogen is classed as a diatomic element as it exists not as single atoms of hydrogen, but as pairs of hydrogen atoms that share electrons in molecules to become stable.

A diatomic molecule is a molecule that contains only two atoms.

There are several other elements that exist as diatomic molecules, all of which you must know. To make it easy to remember all the diatomic elements, use this little mnemonic, or make up your own:

Fancy	Fluorine	F_2
Clancy	Chlorine	Cl_2
Owes	Oxygen	O_2
Him	Hydrogen	H_2
Nothing	Nitrogen	N_2
But	Bromine	Br_2
Ice	Iodine	I_2

For example, oxygen exists as diatomic molecules; that is why its formula is O_2. It is important to remember the names of all seven diatomic elements and to be able to draw a diagram of a molecule of each, showing all the outer electrons, as in the example drawn for hydrogen.

Figure 3.4 shows a molecule of fluorine. An atom of fluorine requires one electron to become stable, so it will share one electron with another atom of fluorine to form two stable atoms in a molecule.

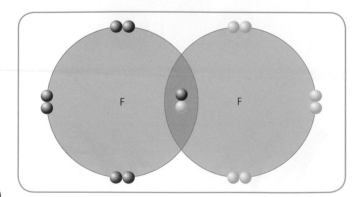

Figure 3.4 Bonding in a diatomic fluorine molecule

Oxygen atoms have the electron arrangement 2,6 and so require two electrons to become stable. Because of this, they must share two electrons with another oxygen atom (Figure 3.5). This means that oxygen forms a double covalent bond.

You may have noticed that so far we have only looked at how the bonds are formed in molecules of elements, but the same rules apply when dealing with the bonding in covalent compounds.

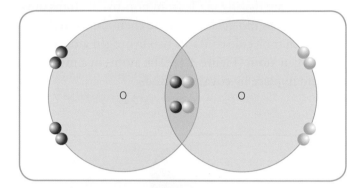

Figure 3.5 Bonding in a diatomic oxygen molecule

Methane (carbon hydride) has the chemical formula CH_4. This formula results from the atoms of carbon and hydrogen sharing electrons to become stable. An atom of carbon has four outer electrons (Figure 3.6) and therefore requires a further four to achieve a stable electron arrangement like a noble gas.

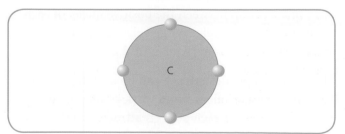

Figure 3.6 A carbon atom

A hydrogen atom has only one electron (Figure 3.7), so it requires only one more electron to achieve a stable electron arrangement.

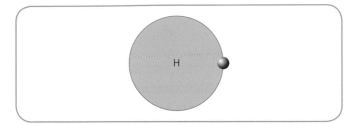

Figure 3.7 A hydrogen atom

For the two to combine and form a stable compound in which both hydrogen and carbon have stable electron arrangements, the carbon atom requires four hydrogen atoms to supply the four electrons (Figure 3.8).

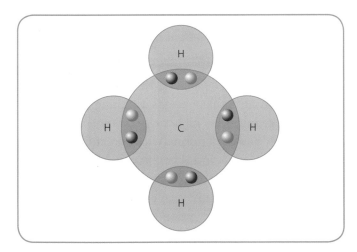

Figure 3.8 Bonding in a methane molecule

Sharing electrons in this way allows both hydrogen and carbon to have stable electron arrangements and this is the reason that methane (carbon hydride) has the formula CH_4.

Shapes of molecules

The repulsion between the pairs of electrons in the bonds results in covalent molecules having distinctive shapes. The bonds will move away from each other to create the shapes of the molecules that follow.

The three-dimensional shapes of molecules can be difficult to draw. To overcome this problem, chemists use the symbols shown in Figure 3.9 when drawing the shapes of molecules.

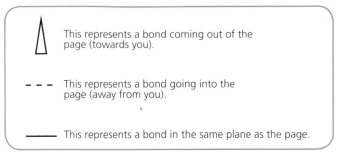

Figure 3.9 Drawing three-dimensional shapes

Tetrahedral molecules

Let's start with the example of methane (CH_4) (Figure 3.10). Many molecules that have a formula like methane, in which one atom of an element is joined to four of another, will have a **tetrahedral** shape. For example, CCl_4 (tetrachloromethane) will also be tetrahedral in shape.

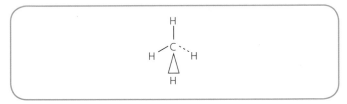

Figure 3.10 The tetrahedral structure of methane

Trigonal pyramidal molecules

If asked to think of a pyramid, most people will picture one of the famous pyramids in Egypt. These structures have a square base. A pyramid with a triangular base is known as a **trigonal pyramid**. Ammonia (NH_3) is an example of a molecule with a trigonal pyramidal shape (Figure 3.11). Many other molecules that have a formula like ammonia, in which one atom of an element is joined to three of another, will have this shape. For example, a molecule of PCl_3 (phosphorus trichloride) will also be trigonal pyramidal in shape.

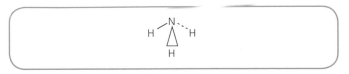

Figure 3.11 The trigonal pyramidal structure of ammonia

Angular molecules

Water (H_2O) is an example of a molecule with an angular shape (Figure 3.12). In molecules containing three atoms, either the atoms form a straight line, in which case the shape is said to be linear, or there is an angle in the shape, like a boomerang.

Linear molecules

Hydrogen fluoride (HF) is an example of a linear molecule (Figure 3.13). All two-atom (diatomic) molecules must be linear.

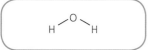

Figure 3.12 The angular structure of water

Figure 3.13 The linear structure of hydrogen fluoride

Properties of covalent substances

Almost all covalent substances, both elements and compounds, are gases, liquids or soft solids with low melting points. Why should this be? Most covalent substances exist as small molecules containing a certain number of atoms held together by covalent bonds. These covalent bonds are very strong. Each molecule exists separately from all other molecules of the same substance. However, in addition to the covalent bonds *within* molecules (intramolecular bonds) which hold the atoms together, bonds also exist *between* molecules (intermolecular bonds). These intermolecular bonds are much weaker than covalent bonds. So when a covalent substance melts, it is only these weak forces of attraction that must be broken (Figure 3.14).

Covalent substances like the ones described above are known as discrete molecular or covalent molecular substances.

Covalent molecular substances tend not to dissolve in water. For example, the polystyrene chips used in packaging simply float if placed in water. However, place them in propanone (a covalent molecular liquid), and they will dissolve. An alternative solvent to water can usually be found for a covalent molecular substance. Covalent molecular compounds never conduct electricity because molecules are not charged particles. Water is an exception to this because a small number of water molecules **dissociate** forming ions.

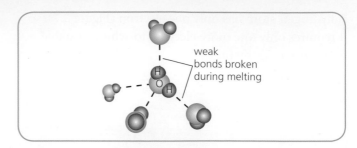

Figure 3.14 Discrete covalent substances melt at low temperatures because the weak bonds between the molecules are easy to break

A few covalent substances have a very different structure, in which all the atoms in a sample are bonded to one another in one enormous molecule, made up of a large number of atoms that would be impossible to count. Since the bonds in these molecules are strong covalent bonds, melting would mean having to break these bonds. As a result, such substances have very high melting points and they are known as **covalent network** substances.

Diamond has a covalent network structure of carbon atoms (Figure 3.15). The many carbon atoms are held together by strong covalent bonds, meaning that in addition to having a very high melting point, diamond is also extremely hard. Interestingly, carbon atoms are also found packed together in a different covalent network structure in a substance called graphite. This different structure gives graphite different properties. For example, graphite is not hard like diamond and, unusually for a non-metal, it conducts electricity. Search the internet to find out how the graphite covalent network structure differs from that of diamond. There are only three covalent network elements (boron, carbon and silicon) and two covalent network compounds (silicon dioxide and silicon carbide) that you will meet in this course.

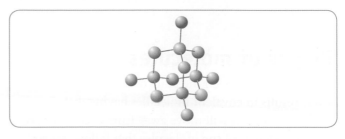

Figure 3.15 The covalent network structure of diamond

As with covalent molecular substances, covalent network substances tend not to dissolve in water, but may dissolve in other solvents.

Ionic bonding

In Chapter 2, we looked at charged particles called ions and the fact that metals atoms form positive ions by losing electrons, and non-metal atoms form negative ions by gaining electrons. The attraction between oppositely charged particles is called an electrostatic attraction.

Ionic bonds are the electrostatic forces of attraction between positive ions and negative ions.

Properties of ionic compounds

Ionic compounds have many different properties. For example, they all have very high melting and boiling points. Why should this be? Ionic compounds have what is known as a lattice structure. This is a regular arrangement of positive and negative ions.

Figure 3.16 shows a diagram of an **ionic lattice**. This ionic lattice has a cube-like structure. If you take a simple grain of salt (sodium chloride) and look at it under a microscope, you will see that the grain is a cube. It is not possible to count the actual numbers of sodium ions and chloride ions in the cube, but you can be sure that the numbers are the same; for every sodium ion there is a chloride ion. This is why the formula for

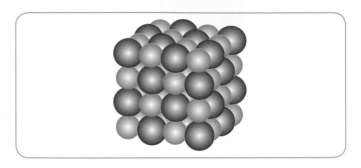

Figure 3.16 An ionic lattice structure

sodium chloride is written as Na^+Cl^-. This represents the ratio of sodium ions to chloride ions in the lattice, in other words 1:1. To melt an ionic compound, millions of strong ionic bonds within the lattice must be broken. Hence, ionic compounds have very high melting points.

Many ionic compounds dissolve easily in water and when they do, their lattice breaks up completely to form free ions. The solutions formed will conduct electricity since ions are charged particles and in solution they are free to move. Moving electrical charges is an electric current. Melting ionic compounds also enables them to conduct electricity by freeing their ions to move.

Conduction can be tested in the laboratory as shown in Figures 3.17, 3.18 and 3.19. In each case, a direct current (d.c.) is used.

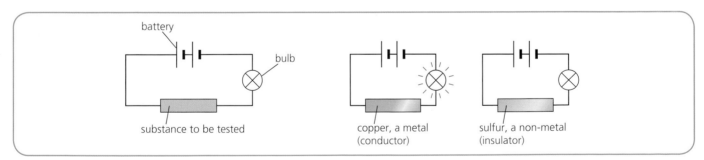

Figure 3.17 Testing a solid substance to see whether it conducts electricity

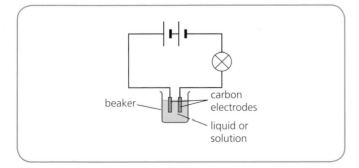

Figure 3.18 Testing a liquid or a solution to see whether it conducts electricity

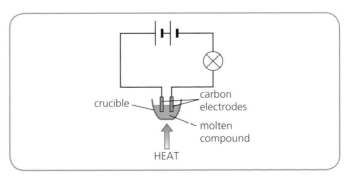

Figure 3.19 Testing a molten compound to see whether it conducts electricity

Summary

Property	Ionic lattice	Covalent network	Covalent molecular
melting and boiling points	high	very high	low
state at room temperature	solid	solid	liquid, gas or low-melting-point solid
conduction of electricity	only when molten or in solution	never (except graphite)	never (although water is a poor conductor)

Table 3.1

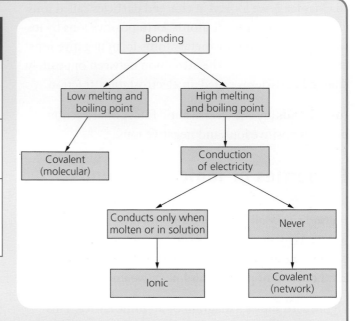

Figure 3.20 Flowchart to summarise types of bonding

Checklist for Revision

- I understand how covalent bonds are formed and the difference between covalent molecular substances and covalent networks.
- I understand how ionic bonds are formed and what ionic lattices are.
- I understand how the physical properties of substances are related to their bonding.

End-of-chapter questions

1 Which of the following substances exists as diatomic molecules?
   A Carbon tetrafluoride
   B Carbon dioxide
   C Calcium oxide
   D Carbon monoxide

2 Which of the following substances does *not* have a covalent network structure?
   A Diamond
   B Sulfur
   C Silicon dioxide
   D Silicon carbide

3 Which of the following has a covalent molecular structure?
   A Neon
   B Sulfur dioxide
   C Sodium chloride
   D Silicon dioxide

4 What type of structure is shown in Figure 3.22?

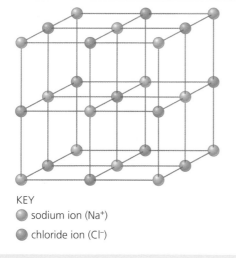

KEY
● sodium ion (Na⁺)
● chloride ion (Cl⁻)

Figure 3.21

   A Covalent network
   B Covalent molecular
   C Ionic lattice
   D Tetrahedral

5 In a molecule of ammonia (NH₃), all the atoms are held together by strong covalent bonds.
   a) Explain fully what a covalent bond is and how a covalent bond holds atoms together.
   b) Draw a diagram of an ammonia molecule showing all the electrons involved in holding the atoms together.
   c) Draw diagrams to show the shapes of methane, ammonia and water molecules.

6 The properties of three different substances are shown in Table 3.2.

Substance	Melting point (°C)	Boiling point (°C)	Conduction
A	−77	−33	no
B	1883	2503	no
C	773	1407	when molten

Table 3.2

Copy and complete Table 3.3 using the letters to show the type of bonding present in each substance.

Substance	Bonding and structure
	ionic
	covalent network
	covalent molecular

Table 3.3

7 Although both contain covalent bonding, carbon dioxide is a gas at room temperature but silicon dioxide is a solid at room temperature. Explain why there is a large difference in melting points between the two covalent compounds.

8 Salt solution conducts electricity but glucose solution does not. What does this suggest about the types of bonding in salt and glucose? Explain your answer.

4   Formulae and reacting quantities

Chemical formulae are an easy way of passing on a lot of information about chemical compounds. Most non-chemists will recognise H_2O as the chemical formula for water, but how many understand what it is telling us about the compound? Few could tell us how to work out the formula for a compound. This section will give you all the tools that you need to tackle chemical formulae.

Names of compounds

The name of a chemical compound gives us important information about the elements that it contains. As a rule, compounds with names ending in -ide contain the two elements indicated by the name. So phosphorus chloride contains the elements phosphorus and chlorine while hydrogen fluoride contains hydrogen and fluorine. The one exception you will regularly encounter to this rule in National 5 is hydroxide compounds such as sodium hydroxide. We will look at these compounds later in Table 4.2.

Compounds with names ending in -ate or -ite contain the elements indicated by the name **and** oxygen. So, while calcium sulfide contains only calcium and sulfur, calcium sulfate and calcium sulfite both contain calcium, sulfur and oxygen. The difference between the sulfate and sulfite compounds can be shown by looking at their chemical formulae, $CaSO_4$ (sulfate) and $CaSO_3$ (sulfite).

Questions

1   Identify the names of the elements present in the following chemical compounds:
    a) magnesium oxide
    b) beryllium bromide
    c) potassium permanganate
    d) lead nitrate
    e) sodium hydrogencarbonate.

What does a formula tell us?

The formula of a covalent molecular compound gives the number of atoms present in a single molecule. For example, the formula for sulfur dioxide (SO_2) tells us that one molecule of the compound contains one sulfur atom and two oxygen atoms.

In ionic and covalent network compounds, the formula tells us something slightly different. The formulae for sodium chloride (NaCl) and silicon dioxide (SiO_2) give the simplest ratio of ions in the sodium chloride lattice (1:1) or atoms in the silicon dioxide network (1:2).

Questions

2   Identify how many of each type of atom are present in one molecule of the following covalent molecular compounds from their formulae.
    a) HF   b) SO_3 c) N_2O_4
    d) PCl_3 e) H_2O

Formulae from names of compounds

Sometimes the name of the compound gives information about its chemical formula. The prefix given to either part of the compound's name gives the number of atoms of that element present in each molecule, as shown in Table 4.1.

Prefix	Number of atoms
mono-	one
di-	two
tri-	three
tetra-	four
penta-	five
hexa-	six

Table 4.1

Examples of chemical names that tell us their formula include carbon tetrachloride, which contains one carbon atom joined to four chlorine atoms, so it has the formula CCl_4. Similarly, water (H_2O) could also be known as dihydrogen monoxide!

Questions

3 Write the chemical formulae for the following chemical compounds:
   a) carbon monoxide
   b) boron tribromide
   c) phosphorus pentachloride
   d) dinitrogen trioxide
   e) xenon hexafluoride.

Valency

The **valency** of an element indicates its combining power, that is, the number of bonds that one of its atoms can form with others. As you have learned, atoms form chemical bonds to fill their outer energy levels of electrons, and so elements in the same group of the Periodic Table have the same valency (see Figure 4.1).

The noble gases have a valency of zero as they do not usually form bonds with other elements.

SVSDF system

The SVSDF system is a set of simple steps to follow which will help you to write the chemical formula for a compound. The letters stand for **S**ymbols, **V**alency, **S**wap, **D**ivide, **F**ormula.

Step 1: Write down the symbols of both the elements involved.
Step 2: Beneath each symbol, write its valency.
Step 3: Swap the valencies over.
Step 4: If the valencies can be simplified, divide them both by the smaller of the two numbers. If one of the numbers is already one, then they cannot be divided and simplified any further.
Step 5: Write the formula.

Group 1 Valency 1	Group 2 Valency 2	Group 3 Valency 3	Group 4 Valency 4	Group 5 Valency 3	Group 6 Valency 2	Group 7 Valency 1	Group 0 Valency 0
1 Hydrogen H 1							2 Helium He 2
3 Lithium Li 2,1	4 Beryllium Be 2,2	5 Boron B 2,3	6 Carbon C 2,4	7 Nitrogen N 2,5	8 Oxygen O 2,6	9 Fluorine F 2,7	10 Neon Ne 2,8
11 Sodium Na 2,8,1	12 Magnesium Mg 2,8,2	13 Aluminium Al 2,8,3	14 Silicon Si 2,8,4	15 Phosphorus P 2,8,5	16 Sulfur S 2,8,6	17 Chlorine Cl 2,8,7	18 Argon Ar 2,8,8
19 Potassium K 2,8,8,1	20 Calcium Ca 2,8,8,2	31 Gallium Ga 2,8,18,3	32 Germanium Ge 2,8,18,4	33 Arsenic As 2,8,18,5	34 Selenium Se 2,8,18,6	35 Bromine Br 2,8,18,7	36 Krypton Kr 2,8,18,8
37 Rubidium Rb 2,8,18,8,1	38 Strontium Sr 2,8,18,8,2	49 Indium In 2,8,18,18,3	50 Tin Sn 2,8,18,18,4	51 Antimony Sb 2,8,18,18,5	52 Tellurium Te 2,8,18,18,6	53 Iodine I 2,8,18,18,7	54 Xenon Xe 2,8,18,18,8
55 Caesium Cs 2,8,18,18,8,1	56 Barium Ba 2,8,18,18,8,2	81 Thallium Tl 2,8,18,32,18,3	82 Lead Pb 2,8,18,32,18,4	83 Bismuth Bi 2,8,18,32,18,5	84 Polonium Po 2,8,18,32,18,6	85 Astatine At 2,8,18,32,18,7	86 Radon Rn 2,8,18,32,18,8
87 Francium Fr 2,8,18,32,18,8,1	88 Radium Ra 2,8,18,32,18,8,2						

Figure 4.1 A section of the Periodic Table showing the valencies of the various groups

Worked examples

1   What is the formula for potassium oxide?

Symbols	K	O
Valency	1	2
Swap	2	1
Divide	2	1
Formula		K_2O

2   What is the formula for aluminium oxide?

Symbols	Al	O
Valency	3	2
Swap	2	3
Divide	2	3
Formula		Al_2O_3

No division has been necessary in these first two examples, because there is no common factor between the two numbers. In other words, they cannot be simplified further.

3   What is the formula for carbon sulfide?

Symbols	C	S
Valency	4	2
Swap	2	4
Divide	$\frac{2}{2}$	$\frac{4}{2}$
	1	2
Formula		CS_2

Remember that the valency of an atom is related to its group number in the Periodic Table.

Questions

4   Using the SVSDF method, write the chemical formulae for the following compounds:
   a) lithium sulfide
   b) potassium chloride
   c) calcium phosphide
   d) boron bromide
   e) magnesium nitride.

Formulae of compounds containing group ions

Group ions contain two or more atoms, and most of them have a negative charge. The formulae of some common group ions are shown in Table 4.2. A similar table can be found in the SQA Data Booklet.

The number of charges on a group ion can be used as the valency of the ion to help with working out formulae.

One positive		One negative		Two negative		Three negative	
Ion	**Formula**	**Ion**	**Formula**	**Ion**	**Formula**	**Ion**	**Formula**
ammonium	NH_4^+	ethanoate	CH_3COO^-	carbonate	CO_3^{2-}	phosphate	PO_4^{3-}
		hydrogencarbonate	HCO_3^-	chromate	CrO_4^{2-}		
		hydrogensulfate	HSO_4^-	dichromate	$Cr_2O_7^{2-}$		
		hydrogensulfite	HSO_3^-	sulfate	SO_4^{2-}		
		hydroxide	OH^-	sulfite	SO_3^{2-}		
		nitrate	NO_3^-				
		permanganate	MnO_4^-				

Table 4.2

Worked example

What is the formula for magnesium nitrate?

Since nitrate is in the 'one negative' column of the group ion table, it has a valency of one.

Symbols	Mg	NO_3
Valency	2	1
Swap	1	2
Divide	1	2
Formula		$Mg(NO_3)_2$

When a formula contains more than one group ion, the group ion *must* be written in brackets. In magnesium nitrate, the small '2' indicates that there are two nitrate ions (two NO_3^- groups) present in the formula.

With ionic compounds, the formula shows us the simplest ratio of each type of ion in the substance. We can take the formula one stage further and write ionic formulae. As you know, an ionic compound is made up of oppositely charged ions (usually a metal ion and a non-metal ion, or a metal ion and a group ion). The ionic formula includes the charges present on each of the ions. The charges are the same as the valency for the element or group ion, remembering of course that all metals will form positively charged ions and non-metals form negatively charged ions.

Worked example

Write the ionic formula of beryllium hydroxide.

Symbols	Be	OH
Valency	2	1
Swap	1	2
Divide	1	2
Formula		$Be(OH)_2$

Beryllium is a group two metal and has a valency of two. This means that a beryllium ion has a 2+ charge. The hydroxide ion is negatively charged and carries a 1− charge.

Ionic formula: $Be^{2+}(OH^-)_2$

The other ionic compounds for which we have worked out formulae in this section are shown in Table 4.3, along with their names, formulae and ionic formulae.

Name	Formula	Ionic formula
potassium oxide	K_2O	$(K^+)_2O^{2-}$
aluminium oxide	Al_2O_3	$(Al^{3+})_2(O^{2-})_3$
magnesium nitrate	$Mg(NO_3)_2$	$Mg^{2+}(NO_3^-)_2$

Table 4.3

Questions

5 Using the formulae for group ions from Table 4.2, write the chemical formulae for the following compounds:
   a) magnesium hydroxide
   b) sodium hydrogencarbonate
   c) potassium permanganate
   d) aluminium sulfate
   e) ammonium phosphate.
6 Write the ionic formulae for each of the compounds in Question 5.

Formulae using Roman numerals

In some cases, particularly with transition metals, the valency of an element is not always the same when it appears in different compounds. This is indicated by showing the valency as a Roman numeral in brackets after the element's name (Table 4.4).

Roman numeral	Number
I	one
II	two
III	three
IV	four
V	five
VI	six

Table 4.4

What is the formula for lead(IV) sulfide?

Symbols	Pb	S
Valency	4	2
Swap	2	4
Divide	$\frac{2}{2}$	$\frac{4}{2}$
	1	2
Formula	PbS_2	

Questions

7 Write the chemical formulae for the following chemical compounds:
  a) gold(III) oxide
  b) tin(IV) chloride
  c) iron(III) sulfide
  d) copper(I) bromide
  e) lead(II) iodide.

Balanced equations

A chemical equation is an easy way of describing a chemical change, showing both the reactants and products. These equations are balanced when there is the same number of each type of atom on both sides of the equation.

Equations can also indicate the state of each substance involved in a chemical reaction. State symbols, which are placed in brackets after the name or formula of each reactant or product, tell you the physical state (solid, liquid, gas or dissolved in water to form an aqueous solution) in which each chemical is found. These state symbols are shown in Table 4.5 below.

State symbol	Physical state
(s)	solid
(l)	liquid
(g)	gas
(aq)	aqueous solution

Table 4.5

As an example, consider what happens when solid calcium granules are burned. Calcium atoms and gaseous oxygen molecules react together to make solid calcium oxide.

The word equation for the reaction would be:

$$\text{calcium} + \text{oxygen} \longrightarrow \text{calcium oxide}$$

The reactants, calcium and oxygen, are written to the left of the arrow and the product, calcium oxide, to the right.

If we simply replace each of the chemicals' names with their chemical symbols or formulae and their state symbols, the equation will be unbalanced:

$$Ca(s) + O_2(g) \longrightarrow CaO(s)$$

Notice that there are unequal numbers of oxygen atoms on the left-hand side (two) compared with the right-hand side (one). To make things equal, you need to adjust the number of formula units of some of the substances until you get equal numbers of each type of atom on both sides. So:

$$Ca(s) + O_2(g) \longrightarrow 2CaO(s)$$

Now there are two oxygen atoms on each side of the equation. However, there are now two calcium atoms on the right-hand side, but only one on the left-hand side. So:

$$2Ca(s) + O_2(g) \longrightarrow 2CaO(s)$$

This is the balanced symbol equation.

Calcium oxide exists as an ionic lattice. Figure 4.2 shows a simplified version of its formation to help you understand the equation balancing process.

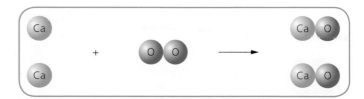

Figure 4.2

In the original equation, there were uneven numbers of oxygen atoms on the left-hand side and right-hand side of the equation. Balancing the oxygen involves adding a two to the right-hand side. Doubling the number of units of calcium oxide produced means that we have to double the quantity of calcium on the reactant side of the equation too.

To balance equations, follow these simple rules:

Step 1: Check that all the formulae in the equation are correct.

Step 2: Deal with only one element at a time.

Step 3: Balancing involves writing numbers *in front of* formulae. You cannot change any of the smaller numbers *within* a chemical formula as this would change the formula.

Step 4: Check each element again; repeat step 3 if needed.

Worked example

Heptane (C_7H_{16}) burns in a plentiful supply of oxygen to produce carbon dioxide and water.

$$C_7H_{16}(l) + O_2(g) \longrightarrow CO_2(g) + H_2O(l)$$

This equation is unbalanced. There are seven carbon atoms on the left-hand side and only one on the right. To balance the carbon, write a '7' in front of the carbon dioxide.

$$C_7H_{16}(l) + O_2(g) \longrightarrow \mathbf{7}CO_2(g) + H_2O(l)$$

We now need to balance the hydrogen. We have 16 atoms of hydrogen on the left-hand side in C_7H_{16} and only two on the right. To balance the hydrogen atoms, write an '8' in front of H_2O.

$$C_7H_{16}(l) + O_2(g) \longrightarrow 7CO_2(g) + \mathbf{8}H_2O(l)$$

Nearly there! Now that the carbon and hydrogen atoms from heptane have been balanced, we are only left with oxygen still to be balanced. We have two atoms of oxygen on the left-hand side, but in total on the right-hand side we have 22 oxygen atoms (fourteen from $7CO_2$ and eight from $8H_2O$). This can be balanced by writing an '11' in front of the diatomic oxygen molecule on the left-hand side.

The balanced equation will be:

$$C_7H_{16}(l) + 11O_2(g) \longrightarrow 7CO_2(g) + 8H_2O(l)$$

Questions

8 Balance the following chemical equations:

a) $C_2H_4(g) + O_2(g) \longrightarrow CO_2(g) + H_2O(l)$

b) $Ag_2S(aq) + Al(s) \longrightarrow Al_2S_3(aq) + Ag(s)$

c) $C_6H_{12}O_6(s) + O_2(g) \longrightarrow CO_2(g) + H_2O(l)$

d) $H_3PO_4(aq) + Mg(OH)_2(s) \longrightarrow Mg_3(PO_4)_2(s) + H_2O(l)$

e) $CO(g) + O_2(g) \longrightarrow CO_2(g)$

The mole

Moles are small, furry mammals that live underground, famous for their burrowing skills and supposedly poor eyesight – but to a chemist, a **mole** is a unit of measurement.

When you try to balance equations, you possibly think in terms of counting the numbers of atoms of each element present on both sides. For example, when presented with:

$$Ca(s) + HCl(aq) \longrightarrow CaCl_2(aq) + H_2(g)$$

you might say to yourself: 'there are two atoms of chlorine and two atoms of hydrogen on the right-hand side but only one of each on the left-hand side'. You may then write the balanced equation:

$$Ca(s) + 2HCl(aq) \longrightarrow CaCl_2(aq) + H_2(g)$$

Once an equation has been balanced successfully, you can use it either to calculate the quantities of reactants needed for a reaction or the quantities of products produced. When it comes to doing the actual experiment, however, you do not count atoms. You could not start off with one atom of calcium in the above example! Enter the mole.

For most elements, 1 mole is the relative atomic mass (RAM) in grams, for example 12 g of carbon or 40 g of calcium. However, for the seven diatomic elements, 1 mole is *twice* the RAM in grams, for example 2 g of hydrogen or 32 g of oxygen. For compounds, we use the **gram formula mass (GFM)**. The gram formula mass is the mass of 1 mole of a compound in grams and is

Figure 4.3 Is this what chemists mean by a mole?

calculated using the RAMs of each of the elements in its formula. For example, 1 mole of water (H_2O) weighs 18 g. This is calculated by adding $(1 + 1 + 16)$ because water molecules contain two hydrogen atoms and one oxygen atom.

know the number of moles of the compound and have calculated the formula mass.

Using the formula triangle is straightforward. Cover whichever of the three variables you are trying to calculate and whatever is left will show you the equation to use. While the formula triangle does not appear in the SQA Data Booklet, it is represented by the formula $n = m/$GFM in the formula list.

Questions

9 Calculate the mass of 1 mole of each of the following compounds:
   a) sulfur dioxide (SO_2)
   b) carbon hydride (CH_4)
   c) potassium permanganate ($KMnO_4$)
   d) calcium carbonate ($CaCO_3$)
   e) magnesium nitrate ($Mg(NO_3)_2$).

Mole calculations

Once you have mastered calculating the formula mass of any compound, you can calculate the mass of a precise number of moles using a formula triangle (Figure 4.4). This formula triangle can also be used to calculate the mass of a compound if you already

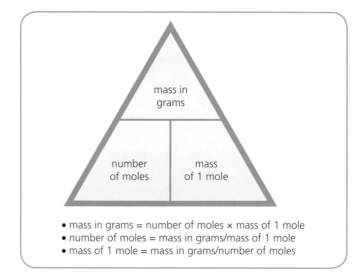

- mass in grams = number of moles × mass of 1 mole
- number of moles = mass in grams/mass of 1 mole
- mass of 1 mole = mass in grams/number of moles

Figure 4.4 The formula triangle

Worked examples

1 What is the mass of 1 mole of calcium chloride?

To calculate this all you have to do is work out the formula of calcium chloride using the SVSDF method, then add together the relative atomic masses of each element.

S	Ca		Cl		$CaCl_2$
V	2		1		
S	1		2		40 g + (2 × 35.5 g)
D	1		2		= 40 + 71
F		$CaCl_2$			= 111 g

Note that since there are two chloride ions in the formula, we must take that into account when calculating the mass.

1 mole of calcium chloride has a mass of 111 g.

2 What is the mass of 1 mole of ammonium phosphate?

S	NH_4		PO_4		$(NH_4)_3PO_4$
V	1		3		
S	3		1		(3 × 14 g) + (12 × 1 g) + 31 g + (4 × 16 g)
D	3		1		= 42 + 12 + 31 + 64
F		$(NH_4)_3PO_4$			= 149 g

1 mole of ammonium phosphate has a mass of 149 g.

Worked examples

1 How many moles are present in 25 g of calcium carbonate?

The formula we need to use, given by covering up number of moles, is:

$$\text{number of moles} = \frac{\text{mass in grams}}{\text{mass of 1 mole}}$$

First, calculate the gram formula mass of calcium carbonate following the method used earlier.

The formula of calcium carbonate is $CaCO_3$.

$Ca = 40\,g$, $C = 12\,g$ and $O = (3 \times 16\,g)$

$GFM = 40 + 12 + 48 = 100\,g$

mass (from the question) = 25 g

$$\text{number of moles} = \frac{\text{mass in grams}}{\text{mass of 1 mole}}$$

$$= \frac{25\,g}{100\,g}$$

$$= 0.25 \text{ moles}$$

2 Calculate the mass of 2.5 moles of sodium nitrate (GFM = 85 g).

Using the triangle to calculate mass gives the formula:

$$\text{mass in grams} = \text{number of moles} \times \text{mass of 1 mole}$$

$$= 2.5 \times 85\,g$$

$$= 212.5\,g$$

Questions

10 Calculate the mass of the following:
  a) 3 moles of sodium oxide (Na_2O)
  b) 1.5 moles of potassium hydroxide (KOH)
  c) 0.25 moles of nitrogen dioxide (NO_2)
  d) 2.5 moles of sulfuric acid (H_2SO_4)
  e) 5 moles of magnesium chloride ($MgCl_2$).

11 How many moles are present in each of the following:
  a) 48 g of methane (CH_4)
  b) 150 g of calcium carbonate ($CaCO_3$)
  c) 6.4 g of sulfur dioxide (SO_2)
  d) 42.5 g of ammonia (NH_3)
  e) 146 g of hydrogen chloride (HCl)?

Calculations from balanced equations

A balanced chemical equation provides information about the reactants, how much you need to use (the number of moles), the products that will be made and how much you can expect to make (also in moles). For example, the equation:

$$Ca + 2HCl \longrightarrow CaCl_2 + H_2$$

tells us that 1 mole of calcium and 2 moles of hydrochloric acid can react to form 1 mole of calcium chloride and 1 mole of hydrogen.

1 mole of calcium weighs 40 g and 1 mole of calcium chloride weighs 111 g, so if you had 40 g of calcium, you could make 111 g of calcium chloride. But what if you only had 20 g of calcium – how much calcium chloride could you make? You can probably see that using only half as much calcium will lead to making half as much calcium chloride – 55.5 g. But what if you had, say, 8 g of calcium? Take a look at the step-by-step example below, then try Question 12 for yourself.

Step 1: Write a balanced chemical equation, unless already given.

$$Ca + 2HCl \longrightarrow CaCl_2 + H_2$$

Step 2: Identify the two chemicals referred to in the question and write the mole ratio

1 mole $\longrightarrow$ 1 mole

Step 3: Calculate the number of moles of the substance you have been given a mass for (Ca)

$$\text{number of moles} = \frac{\text{mass in grams}}{\text{mass of 1 mole}}$$

$$= \frac{8\,g}{40\,g}$$

$$= 0.2 \text{ mole}$$

Step 4: Use the mole ratio to calculate the number of moles of the substance you are trying to find (CaCl₂)

1 mole $\longrightarrow$ 1 mole

0.2 mole $\longrightarrow$ 0.2 mole

Step 5: Calculate the mass

mass in grams = number of moles $\times$ mass of 1 mole

$$= 0.2 \times 111\,g$$

$$= 22.2\,g$$

Questions

12 In each of the following problems, a balanced equation is provided. You will not always be so lucky!

a) Calculate the mass of hydrogen produced when 5 g of calcium reacts with an excess of dilute sulfuric acid.

$$Ca + H_2SO_4 \longrightarrow CaSO_4 + H_2$$

b) Calculate the mass of iron produced from 2 g of iron(III) oxide in the following reaction:

$$2Al + Fe_2O_3 \longrightarrow 2Fe + Al_2O_3$$

c) Calculate the mass of carbon dioxide formed when 0.8 g of methane is completely burned.

$$CH_4 + 2O_2 \longrightarrow CO_2 + 2H_2O$$

d) Calculate the mass of water produced by burning 10 g of hydrogen.

$$2H_2 + O_2 \longrightarrow 2H_2O$$

Mole calculations involving solutions

Many chemical reactions are not carried out using solids or liquids, quantities of which can easily be weighed. Some are carried out using **solutions**. A solution is formed when a **solute** is dissolved in a **solvent**. The **concentration** of a solution is measured in moles per litre ($mol\,l^{-1}$). Now that you know what a mole is,

perhaps this unit for concentration makes more sense. If you have 1 litre of a solution and it contains 1 mole of solute, then the concentration of the solution is obviously 1 mole per litre. Imagine you took a 500 cm³ sample from this solution. What would the concentration of your sample be? The answer is still 1 mole per litre, because you will have 0.5 of a mole of solute in 0.5 of a litre of solution. The concentration of any solution can be calculated using the formula triangle given in Figure 4.5.

The volume used must be in litres. To convert cm³ to litres, simply divide by 1000. For example, if the volume is given as 250 cm³, then in litres this is 0.25 l (250/1000 = 0.25).

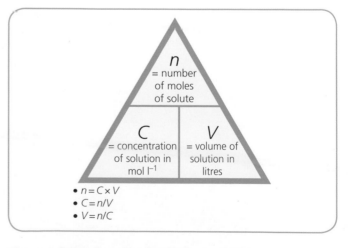

- $n = C \times V$
- $C = n/V$
- $V = n/C$

Figure 4.5 The formula triangle for a solution

Worked example

Calculate the concentration of a potassium nitrate solution if 3 moles are dissolved in 750 cm³ of solution.

As before, covering concentration in our triangle gives the formula we want to use:

$$C = \frac{n}{V}$$

$n = 3$ moles

$V = 750 \, cm^3 = 0.75 \, l$

$$C = \frac{n}{V}$$

$$= \frac{3}{0.75}$$

$$= 4 \, mol \, l^{-1}$$

Questions

13 Calculate the concentration of the following solutions in $mol \, l^{-1}$:
   a) 0.25 moles of sodium hydroxide dissolved in 500 cm³ of solution
   b) 3 moles of magnesium chloride dissolved in 2 litres of solution
   c) 1.5 moles of nitric acid dissolved in 1500 cm³ of solution
   d) 6 moles of potassium bromide dissolved in 3 litres of solution
   e) 0.6 moles of potassium iodide dissolved in 250 cm³ of solution.

More complicated examples

Sometimes you will meet a question that cannot be answered by either formula triangle individually. These problems must be solved using *both* of the triangles that we have learned about in one calculation.

Worked examples

1 **What mass of calcium hydroxide is required to produce 500 cm³ of 0.1 mol l⁻¹ calcium hydroxide solution?**

The question gives us a *concentration* and a *volume*, but these cannot be used alone to give us the required mass.

First, we must calculate the number of moles of calcium hydroxide that are required.

$n = ?$

$C = 0.1 \, mol \, l^{-1}$

$V = 500 \, cm^3 = 0.5 \, l$

$n = C \times V$

$= 0.1 \times 0.5$

$= 0.05$ moles

The number of moles can be substituted into the other formula triangle and used to calculate the mass of calcium hydroxide required.

Remember that even if you are not given the formula mass, it can be calculated from the formula using the relative atomic masses of each element.

$Ca(OH)_2$

$Ca = 40, O = (2 \times 16), H = (2 \times 1)$

$= 40 + 32 + 2$

$= 74 \, g$

mass in grams = number of moles $\times$ mass of 1 mole

$= 0.05 \times 74 \, g$

$= 3.7 \, g$

2 **What is the concentration of a solution containing 5 g of sodium hydroxide in 250 cm³ of solution?**

mass = 5 g

number of moles = ?

formula mass of NaOH = 23 + 16 + 1

$= 40$

number of moles of NaOH = $\dfrac{mass \, in \, grams}{mass \, of \, 1 \, mole}$

$= \dfrac{5 \, g}{40 \, g}$

$= 0.125$ moles

Calculating the number of moles means that we can use $C = n/V$ to calculate the concentration.

$n = 0.125$ (calculated using the other triangle)

$C = ?$

$V = 250 \, cm^3$ (given in the question) $= 0.25 \, l$

concentration of NaOH = $\dfrac{n}{V}$

$= \dfrac{0.125}{0.25}$

$= 0.5 \, mol \, l^{-1}$

Questions

14 Calculate the concentration of the following solutions in $mol\,l^{-1}$:
  a) 10.1 g of potassium nitrate (KNO_3) dissolved in 500 cm^3 of solution
  b) 5.55 g of calcium chloride ($CaCl_2$) dissolved in 50 cm^3 of solution
  c) 4 g of sodium hydroxide (NaOH) dissolved in 250 cm^3 of solution
  d) 69 g of potassium carbonate (K_2CO_3) dissolved in 250 cm^3 of solution
  e) 1.49 g of ammonium phosphate (($NH_4)_3PO_4$) dissolved in 20 cm^3 of solution.

Percentage composition

A chemical formula tells us the number of atoms in a molecule of a covalent compound or the ratio of ions in an ionic compound. Percentage composition is the percentage by mass of each element present. It can be calculated using the formula:

$$\% \text{ by mass} = m/\text{GFM} \times 100$$

In this formula, '% by mass' represents the percentage of the total mass of the compound that a particular element accounts for; 'm' represents the mass of the element that is present in the compound in grams. Be careful with this value – it may simply be the relative atomic mass of the element in grams, but can be more if the formula shows that more than one atom or ion of a particular element is present. GFM represents the gram formula mass.

Worked examples

1 Calculate the percentage by mass of oxygen in 1 mole of water (H_2O).
   GFM of H_2O = $(2 \times 1\,g) + 16\,g$
   $= 2\,g + 16\,g$
   $= 18\,g$

   $\% \text{ by mass of O in } H_2O = \dfrac{\text{mass of O in } H_2O}{\text{GFM of } H_2O} \times 100$

   $= \dfrac{16}{18} \times 100$

   $= 88.89\%$

2 Calculate the percentage by mass of iron in iron(III) oxide (Fe_2O_3).
   GFM of Fe_2O_3 = $(2 \times 56\,g) + (3 \times 16\,g)$
   $= 112\,g + 48\,g$
   $= 160\,g$

   $\% \text{ by mass of Fe in } Fe_2O_3 = \dfrac{\text{mass of Fe in } Fe_2O_3}{\text{GFM of } Fe_2O_3} \times 100$

   $= \dfrac{(2 \times 56)}{160} \times 100$

   $= 70\%$

Questions

15 Calculate the percentage by mass of the element indicated in each of the following chemical compounds:
  a) carbon in methane (CH_4)
  b) sulfur in sulfur trioxide (SO_3)
  c) oxygen in calcium carbonate ($CaCO_3$)
  d) sodium in sodium sulfate (Na_2SO_4)
  e) hydrogen in butane (C_4H_{10}).

Checklist for Revision

- I can write chemical and ionic formulae, including formulae containing group ions.
- I can write balanced chemical equations.
- I can perform calculations based on balanced chemical equations.
- I can calculate the percentage mass composition of compounds.

5   Acids and bases

You will know about the **pH** scale either from lower school science lessons or from National 4 Chemistry. The pH of a substance is much more than a simple number. It gives important information about how the substance can be expected to react. In this topic, you will learn all about **acids** and their chemical opposites, **bases**. Many of the products you find in cupboards at home are acids or bases (Figure 5.1).

Figure 5.1 a) Household items containing acids

Figure 5.1 b) Household items containing bases

The pH scale

How acidic or basic a substance is (the pH of the substance) can be measured using the pH scale (Figure 5.2), a continuous range that stretches from below 0 to above 14. Most common pH values occur between 0 and 14.

- Solutions of acids have a pH of less than 7.
- Water and neutral solutions have a pH of exactly 7.
- Solutions of bases have a pH of more than 7.

The pH scale is actually a measure of the concentration of hydrogen ions in any given solution. The pH can be measured easily using a variety of pH **indicators**. Some, such as universal indicator, will show a colour for every pH value.

Other, more specialist, indicators are only used over a shorter range of pH and show fewer colour changes. These indicators are more useful for observing chemical reactions as they involve a sharper colour change. A summary of some common indicators is included in Table 5.1.

Indicator	Colours with pH changes
phenolphthalein	colourless below 8.3, pink above 10.0
methyl orange	red below 3.1, yellow above 4.4
thymolphthalein	colourless below 9.2, blue above 10.5
bromothymol blue	yellow below 6.0, blue above 7.6

Table 5.1

The pH of solutions

When we measure the concentration of hydrogen ions in a solution, we discover that acidic, alkaline (an **alkali** is a soluble base) and neutral solutions contain different concentrations of different ions. The important ions when dealing with pH are hydrogen ions (H^+) and hydroxide ions (OH^-).

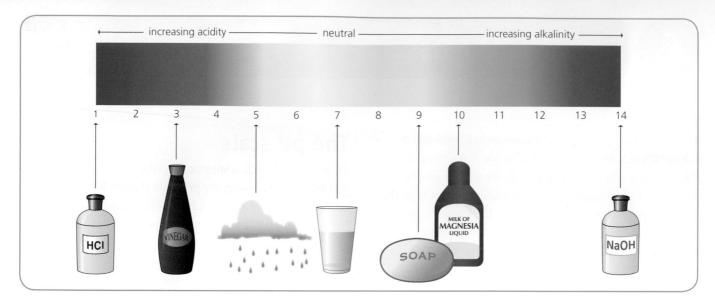

Figure 5.2 The pH scale

Water and neutral solutions

Water molecules can break down into hydrogen and hydroxide ions:

$$H_2O(l) \rightleftharpoons H^+(aq) + OH^-(aq)$$

The arrows in this equation ($\rightleftharpoons$) indicate that this is a **reversible** reaction. This means that the reaction occurs in both the forwards and backwards directions. A small proportion of water molecules dissociate (break up) to form hydrogen ions and hydroxide ions. Some of these hydrogen and hydroxide ions then react together again to form water molecules. A balance is established and the concentration of water molecules, hydrogen ions and hydroxide ions remains constant. At any time only a few water molecules are dissociated into free ions. It is the presence of this small number of ions that explains why water is able to conduct electricity.

In pure water and neutral solutions, the concentration of hydrogen ions is equal to the concentration of hydroxide ions.

Acidic solutions

Acidic solutions have a pH of less than 7. All acids are sources of hydrogen ions and contain H^+ ions in a greater concentration than OH^- ions. For any solution to be acidic, it must have a higher concentration of H^+ ions than OH^- ions.

The H^+ ions present in acidic solutions are responsible for acids reacting in the way they do. Acidic solutions conduct electricity well because they contain oppositely charged ions that are free to move in solution and can carry charge.

Look at the formulae of the common laboratory acids in Table 5.2. When the ionic formulae are written, it is clear that they all contain H^+ ions.

Acid name	Formula	Ionic formula
hydrochloric acid	HCl	$H^+(aq) + Cl^-(aq)$
sulfuric acid	H_2SO_4	$2H^+(aq) + SO_4^{2-}(aq)$
nitric acid	HNO_3	$H^+(aq) + NO_3^-(aq)$
ethanoic acid	CH_3COOH	$H^+(aq) + CH_3COO^-(aq)$

Table 5.2 The formulae of some common laboratory acids

Strongly acidic solutions have a low pH. As acidity decreases, the pH number increases towards 7 on the pH scale.

Alkaline solutions

Alkalis are soluble bases. Alkaline solutions have a pH greater than 7. This is because all alkaline solutions contain a higher concentration of hydroxide (OH⁻) ions than hydrogen ions (H⁺). Like acids, solutions of alkalis also conduct electricity because their ions are free to move in solution and can carry the charge.

The equation below shows what happens to sodium hydroxide when it is dissolved in water:

$$Na^+OH^-(s) \longrightarrow Na^+(aq) + OH^-(aq)$$

Metal hydroxides that are very soluble in water are all strong bases and produce solutions with a high pH when dissolved.

Table 5.3 shows the names and formulae of some common strong laboratory alkalis.

Alkali name	Ionic formula of solid	Ionic formula of solution
sodium hydroxide	Na^+OH^- (s)	$Na^+(aq) + OH^-(aq)$
barium hydroxide	$Ba^{2+}(OH^-)_2$ (s)	$Ba^{2+}(aq) + 2OH^-(aq)$
lithium hydroxide	Li^+OH^- (s)	$Li^+(aq) + OH^-(aq)$

Table 5.3 The formulae of some common laboratory alkalis

Dilution

Diluting an acid makes its pH go up towards 7 and diluting an alkali makes its pH go down towards 7. So adding water to an acid or alkali is one way of changing its pH. But what is actually happening to a solution when water is added? Why does the pH change at all? The answer is that although the number of moles (n) of hydrogen or hydroxide ions has not changed, the volume (V) in which they are dissolved has increased. As a result, the concentration of hydrogen or hydroxide ions has decreased and the pH will change accordingly.

Diluting an acid is not the most effective way of making an acid into a neutral solution, but let us look at how this could be achieved.

To change the pH of an acid by 1, a ten-fold **dilution** is required. If, for example, we begin with 1 cm³ of hydrochloric acid solution of concentration 0.1 mol l⁻¹ and measure its pH, we will find it has a pH of 1. To dilute the acid enough to change its pH to 2, we must add 9 cm³ of water to 1 cm³ of the acid to obtain a ten-fold dilution (1 cm³ + 9 cm³ = 10 cm³). To increase the pH of our 10 cm³ of acid to 3, we would need to add an additional 90 cm³ of water (Figure 5.4). Can you calculate the total volume of water that must be added to make our original 1 cm³ of acidic solution neutral? The data in Table 5.4 should help.

Volume	1 cm³	10 cm³	100 cm³				
pH	1	2	3	4	5	6	7

Table 5.4

Summary

All acidic solutions contain more hydrogen ions than hydroxide ions.

All alkaline solutions contain more hydroxide ions than hydrogen ions.

Neutral solutions contain an equal balance of hydrogen and hydroxide ions.

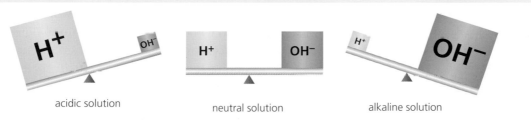

acidic solution   neutral solution   alkaline solution

Figure 5.3 The balance of ions in acidic solutions, neutral solutions and alkaline solutions

Figure 5.4 A 100-fold dilution causes a pH change of 2 units

When an acidic solution is diluted with water, the concentration of H⁺ ions decreases and the pH of the solution increases towards 7. The solution is becoming less acidic. Similarly, when an alkali is diluted with water, the concentration of OH⁻ ions decreases. This causes the pH of the alkali to fall towards 7, making the solution less alkaline as more water is added.

Forming acids and bases

Acids and bases have a wide variety of uses, but how are they obtained?

An alkaline solution can be formed if a soluble metal oxide is dissolved in water. This process follows the equation:

metal oxide + water → metal hydroxide

Burning some magnesium ribbon in a gas jar of oxygen is an exciting experiment that produces a bright, white light (Figure 5.5a). If water and universal indicator are added and the gas jar shaken, the indicator shows us that the product of this **combustion** reaction dissolves in water to form an alkaline solution (Figure 5.5b).

This process can be represented as shown:

magnesium oxide + water ⟶ magnesium hydroxide

$$MgO + H_2O \longrightarrow Mg(OH)_2$$

All other bases which are soluble in water (including the metal oxides of the alkali metals in Group 1 of the Periodic Table) will also dissolve to form solutions with a pH greater than 7. When they dissolve, the soluble

Figure 5.5 Making an alkaline solution of magnesium oxide

metal oxides produce hydroxide ions, causing the pH to increase. Metal oxides outside of Groups 1 and 2 of the Periodic Table are insoluble in water.

Conversely, non-metal oxides, such as sulfur dioxide and nitrogen dioxide, will dissolve in water to form acidic solutions. These two oxides are products of the combustion of many fuels and they contribute to the creation of acid rain.

Summary

Soluble metal oxides produce alkaline solutions when dissolved in water.

Soluble non-metal oxides produce acidic solutions when dissolved in water.

Neutralisation

Neutralisation is the reaction of an acid with a base that results in the pH moving towards 7. It is a useful process that can be applied to situations in everyday life. The treatment of a bee sting (which contains methanoic acid) with baking soda (a common household base) is an example of a neutralisation reaction. Similarly, adding lime (a base) to lakes reduces their acidity and helps to counteract the effects of acid rain.

These reactions can occur between a wide variety of acids and bases, but no matter which acid you choose to neutralise which base (and vice versa), water is always produced as a result of these reactions.

Figure 5.6 The pH lift

Three general word equations that show the products of neutralisation reactions are given below. Remembering these will help you when it comes to working out what is being produced in neutralisation reactions.

acid + metal oxide $\longrightarrow$ salt + water

acid + metal hydroxide $\longrightarrow$ salt + water

acid + metal carbonate $\longrightarrow$ salt + water + carbon dioxide

Basic substances neutralise acids, resulting in the pH of the acid increasing towards 7 and water being produced. (Remember: a soluble base dissolves in water to form an alkaline solution.) Bases include metal oxides, metal hydroxides (alkalis), metal carbonates and ammonia. You will learn about ammonia in Chapter 11.

Naming salts

The salt formed depends on which acid and alkali have been used in the neutralisation reaction. The salt that you add to chips or cook with is called sodium chloride, but there are many other salts. Sodium chloride is readily mined or obtained from seawater but can also be made in the laboratory by a neutralisation reaction.

To name the salt that is produced by any neutralisation reaction, we must appreciate the chemical change that both the acid and the alkali undergo. To form the salt, the metal ion from the alkali (or base) replaces the hydrogen ion from the acid (alkali to front, acid to back).

For example:

hydrochloric + sodium $\longrightarrow$ sodium + water
acid   hydroxide   chloride

HCl + NaOH $\longrightarrow$ NaCl + H_2O

The acid used will determine the type of salt that is produced. Different laboratory acids supply different negative ions to the process and so the name of the acid can be used to tell what the name of the salt will be, as shown in Table 5.5.

Name of acid	Salt name ends with ...
hydrochloric acid	... chloride
sulfuric acid	... sulfate
nitric acid	... nitrate

Table 5.5

The first part of the salt's name comes from the metal present in the base. Some examples are shown in Table 5.6.

Name of base	Salt name starts with ...
sodium hydroxide	sodium ...
potassium hydroxide	potassium ...
magnesium hydroxide	magnesium ...

Table 5.6

So, in a neutralisation reaction between sulfuric acid and potassium hydroxide, the products would be potassium sulfate and water. This can be represented by the following word equation:

sulfuric + potassium $\longrightarrow$ potassium + water
acid   hydroxide   sulfate

H_2SO_4 + 2KOH $\longrightarrow$ K_2SO_4 + $2H_2O$

During neutralisation of an acid with an alkali, the H^+ ion from the acid joins with the OH^- ion from the alkali. This is why water is *always* formed in all such neutralisation reactions:

$$H^+ + OH^- \longrightarrow H_2O$$

Acids and metal carbonates

Acids can also be neutralised by metal carbonates. In these neutralisation reactions, there are three products. The hydrogen ions (H^+) from the acid react with carbonate ions (CO_3^{2-}) to form water and carbon dioxide gas. A salt is also produced. Your stomach contains hydrochloric acid with a pH of between 2 and 3. When you eat too quickly, this acid builds up, leading to acid indigestion. To relieve the discomfort, the extra acid must be neutralised. Common indigestion remedies contain bases such as sodium hydrogencarbonate or calcium carbonate. For example:

hydrochloric + calcium $\longrightarrow$ calcium + water + carbon
acid   carbonate   chloride   dioxide

$$2HCl + CaCO_3 \longrightarrow CaCl_2 + H_2O + CO_2$$

The salt is named in the same way as before, taking the metal from the carbonate and the ending from the acid used.

In the experiment shown in Figure 5.7, we can prove that carbon dioxide is the gas produced by bubbling it through limewater. The limewater will change colour from colourless to cloudy white indicating the presence of carbon dioxide gas.

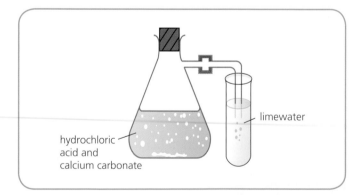

Figure 5.7 Limewater can be used to show that carbon dioxide is released when an acid reacts with a metal carbonate

Ionic equations

Observing the chemical change when an acid and base react together can be difficult to do. We get a more complete picture of the reaction that is taking place when we not only consider the symbol equation for the reaction, but also the ionic equation. Ionic equations are different from symbol equations because they show all the ions involved in the reaction separated out with their charges.

As an example, consider a solution of hydrochloric acid being neutralised by solid copper(II) oxide.

$$2HCl + CuO \rightarrow CuCl_2 + H_2O$$

The salt formed is copper(II) chloride. The ionic equation for the above process, showing all the charges involved, looks like this:

$$2H^+(aq) + 2Cl^-(aq) + Cu^{2+}O^{2-}(s) \rightarrow$$
$$Cu^{2+}(aq) + 2Cl^-(aq) + H_2O(l)$$

Inspection of this equation shows that copper(II) ions and chloride ions appear on both sides; they have not undergone a chemical change (the copper ions have only changed state – a physical change). They are referred to as **spectator ions** (coloured red in the examples which follow). Spectators who attend a football match or a concert do not actually participate. Spectator ions are ions that appear on both the left-hand side and the right-hand side of ionic equations. If we remove the spectator ions, the ionic equation above can be shortened.

$$2H^+(aq) + 2Cl^-(aq) + Cu^{2+}O^{2-}(s) \rightarrow$$
$$Cu^{2+}(aq) + 2Cl^-(aq) + H_2O(l)$$

Rewriting the equation without the spectator ions leaves us with the following chemical change for the neutralisation reaction between an acid and a metal oxide:

$$2H^+(aq) + O^{2-}(s) \rightarrow H_2O(l)$$

Similarly, when an acid is neutralised by a metal hydroxide, the ionic equation can be written and spectator ions removed. Consider a solution of nitric acid being neutralised by lithium hydroxide.

$$HNO_3 + LiOH \rightarrow LiNO_3 + H_2O$$

The salt formed is lithium nitrate. The ionic equation for the above process, showing all the charges involved, looks like this:

$$H^+(aq) + NO_3^-(aq) + Li^+(aq) + OH^-(aq) \rightarrow$$
$$Li^+(aq) + NO_3^-(aq) + H_2O(l)$$

Again, spectator ions that are present on both the left-hand side and the right-hand side of the ionic equation can be identified. Removing them allows a simplified equation to be written:

$$H^+(aq) + NO_3^-(aq) + Li^+(aq) + OH^-(aq) \rightarrow$$
$$Li^+(aq) + NO_3^-(aq) + H_2O(l)$$

The neutralisation equation rewritten with spectator ions omitted is as follows:

$$H^+(aq) + OH^-(aq) \rightarrow H_2O(l)$$

This equation would be the same for any reaction between an acid and a soluble metal hydroxide.

When an acid and metal carbonate react together to form a salt, water and carbon dioxide, the ionic equation without spectator ions also includes the carbon dioxide gas formed. For example, when sulfuric acid is neutralised by sodium carbonate (a soluble metal carbonate), the formulae equation is as follows:

$$H_2SO_4 + Na_2CO_3 \rightarrow Na_2SO_4 + H_2O + CO_2$$

Rewriting this equation as an ionic equation gives:

$$2H^+(aq) + SO_4^{2-}(aq) + 2Na^+(aq) + CO_3^{2-}(aq) \rightarrow$$
$$2Na^+(aq) + SO_4^{2-}(aq) + H_2O(l) + CO_2(g)$$

Note that two covalent substances, water and carbon dioxide, are produced. Again, we can remove the spectator ions and rewrite the simplified equation.

$$2H^+(aq) + SO_4^{2-}(aq) + 2Na^+(aq) + CO_3^{2-}(aq) \rightarrow$$
$$2Na^+(aq) + SO_4^{2-}(aq) + H_2O(l) + CO_2(g)$$

$$2H^+(aq) + CO_3^{2-}(aq) \rightarrow H_2O(l) + CO_2(g)$$

If the same acid was neutralised by an insoluble metal carbonate, such as copper(II) carbonate, the same steps can be repeated to remove spectator ions.

$$2H^+(aq) + SO_4^{2-}(aq) + Cu^{2+}CO_3^{2-}(s) \rightarrow$$
$$Cu^{2+}(aq) + SO_4^{2-}(aq) + H_2O(l) + CO_2(g)$$

$$2H^+(aq) + SO_4^{2-}(aq) + Cu^{2+}CO_3^{2-}(s) \rightarrow$$
$$Cu^{2+}(aq) + SO_4^{2-}(aq) + H_2O(l) + CO_2(g)$$

This results in the chemical change below which can be written for the neutralisation reaction between an acid and an insoluble metal carbonate:

$$2H^+(aq) + CO_3^{2-}(s) \rightarrow H_2O(l) + CO_2(g)$$

Titrations

A **titration** is an experiment in which a *measured* volume of one solution is added to a *known* volume of another solution until the reaction between the two is complete. If the concentration of one of the solutions is known, then the concentration of the other can be calculated. In such experiments, the solution whose concentration is known is referred to as a **standard solution**. In the National 5 course, we use titration to explore reactions between solutions of acids and alkalis.

When titrating, volumes of solutions are measured with great accuracy, using special pieces of glassware called **pipettes** and **burettes**. Unlike a measuring cylinder, a pipette can be used to measure only one precise volume of a solution. Pipettes come in various sizes, but they are all the same shape and each has only a single graduation mark.

In a titration, a pipette is normally used to measure an exact volume of alkali. The pipette is filled with liquid until the bottom of the **meniscus** sits exactly on the graduation mark. The liquid is then transferred into a conical flask. Alkalis are seldom added to burettes because most alkalis are solutions of solids in water and if used in burettes, the tap can clog when the water evaporates if the user forgets to clean the burette thoroughly after use. Acid solutions are usually made by dissolving liquids and gases in water and so this problem does not arise.

Imagine you had a solution of an alkali of unknown concentration and you wanted to find out what the concentration was. Using a pipette (Figure 5.8), you would extract an exact known volume of the alkali and add it to a conical flask, followed by a few drops of a suitable indicator, chosen based on the predicted end-point of the titration. Some of the most common indicators are shown in Table 5.1 on page 35.

The acid solution that you are using to neutralise the alkali in the flask must be of known concentration. This solution is added to a burette. A typical school burette (Figure 5.9) has a scale marked in $0.1\,cm^3$ units, from $0.0\,cm^3$ at the top to $50.0\,cm^3$ at the bottom. This is the reverse of a measuring cylinder, where the scale goes upwards from the bottom! This is because ultimately an experimenter needs to know how much solution has been dispensed *from* the burette, rather than how much is still in it. If you fill one to the zero mark then open the tap to let the solution flow out, when you close the tap, the volume of solution dispensed is simply the new reading on the scale. As with a pipette, the reading

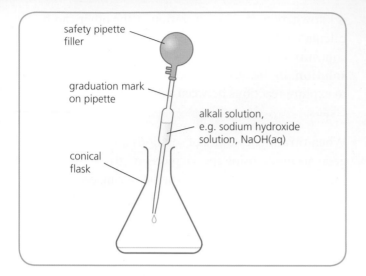

Figure 5.8 Pipetting an exact volume of alkali

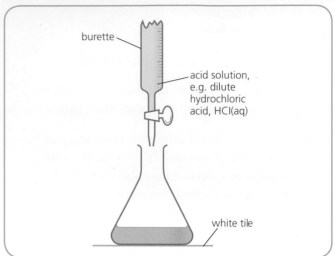

Figure 5.9 Running acid in from a burette

is taken from the bottom of the meniscus against the numerical scale. When titrating, it is not necessary to fill burettes to the zero mark. As long as you note accurately, to one decimal place, what the reading on the scale is before you begin a titration, then note the new reading on the scale once the desired colour change has been observed showing that the titration has been completed, then the *difference* between these two readings will be equal to the volume of solution used. Exactly the same method can be used if you know the concentration of the alkali and want to find out the concentration of the acid.

To ensure accuracy, a rough titration is usually carried out first. This provides an approximate end-point for the reaction, which can be accurately measured through repeatedly carrying out the titration until **concordant** results are obtained.

Titrating with accuracy

Here are a few key points to remember that will help you make accurate titrations:

- The conical flask should be sitting on a white tile. This will help you to observe colour changes more accurately.
- All measurements involving pipettes and burettes should be taken at eye level from the bottom of the meniscus.
- When emptying the pipette, the end should be tapped against the inner wall of the conical flask to ensure it empties completely.

- A rough titration should be carried out first to give an approximate end-point.
- After this, the first accurate titration can be carried out by adding solution from the burette to within 2 or 3 cm³ of the rough titration result, followed by drop by drop addition as the end-point is approached.
- The titration should be repeated until two concordant results (within 0.2 cm³ of each other) are obtained.
- When using the results for calculations, always *ignore* the rough titration result and calculate the average of the two concordant titrations.

Preparation of soluble salts

Titration experiments can be used to obtain pure samples of soluble salts. At the end-point of a neutralisation between an acid and a base, the reaction is exactly complete, with only salt and water in the reaction vessel, and no acid or base remaining. In titrations, you will be able to identify this end-point thanks to the indicator. To obtain a sample of the salt solution that is uncontaminated by the presence of the indicator, the titration should be repeated by adding the known volume required to neutralise the acid **without** the indicator present. Through accurate measurement, you will be left with only a salt solution. To obtain a sample of the dry salt, the water must be removed. To do this, the salt solution can be placed in an evaporating basin and heated with a Bunsen burner (as shown in Figure 5.11) until all of the water has evaporated and only the salt remains.

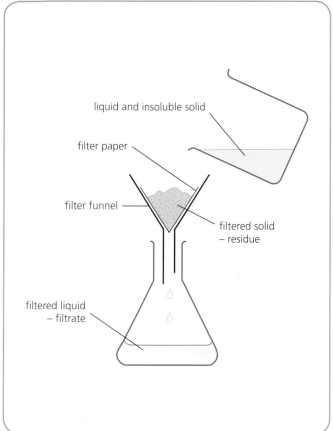

Figure 5.10 Apparatus for filtration

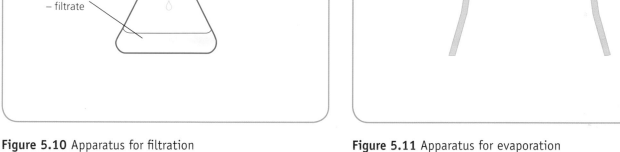

Figure 5.11 Apparatus for evaporation

Soluble salts can also be prepared by the reaction between an acid and an insoluble metal oxide or metal carbonate. An **excess** of either of these bases can be added to the acid to ensure that all of the acid has been neutralised. In the reaction between an acid and a metal carbonate, the metal carbonate powder can be added until no more fizzing occurs, as carbon dioxide gas will only be produced while acid remains. After excess base has been added, a mixture of the desired salt solution and insoluble base will remain. To obtain a pure salt solution, the reaction mixture must be filtered (as shown in Figure 5.10). Filtration involves passing the mixture of insoluble solid and solution through a filter funnel. The insoluble solid (in this case, the excess base) will remain as a residue in the filter paper, while the salt solution can be collected as the filtrate.

The dry salt can then be obtained as before by evaporation (as shown in Figure 5.11).

Calculations based on titrations

There are several different methods that can be used to calculate an unknown concentration of an acid or alkali based on the results of a titration experiment. No matter which method you follow, the results of the calculation will be the same.

Worked example

A student is trying to calculate the concentration of a solution of sodium hydroxide. She titrates $20\,cm^3$ portions of the alkali against a $0.1\,mol\,l^{-1}$ solution of sulfuric acid. Her results are shown in Figure 5.12.

Use the student's results to calculate the concentration of the sodium hydroxide.

A useful starting point, no matter which method you use for the calculation, is a balanced symbol equation for the process. The reaction that takes place is:

$$H_2SO_4 + 2NaOH \longrightarrow Na_2SO_4 + 2H_2O$$

The average titration of the student's experiment can be calculated as the average of her two concordant results. Her average titre is:

$$\frac{(15.0 + 15.0)}{2} = 15.0\,cm^3$$

Using the information known about the acid and the balanced equation, the unknown concentration can be calculated. This method uses the formula triangle that you learned earlier and the balanced equation.

Balanced equation:

$$H_2SO_4 + 2NaOH \longrightarrow Na_2SO_4 + 2H_2O$$

    1 mole   2 moles

The first job is to calculate the number of moles of acid that reacted:

number of moles $H_2SO_4 = C \times V$

$$= 0.1 \times 0.015$$

$$= 0.0015$$

From the equation: 1 mole of H_2SO_4 would react with 2 moles of NaOH. We have calculated that only 0.0015 moles of acid were present, therefore double that number (0.003 moles) of NaOH were neutralised.

Results			
Titration	Start volume (cm³)	End volume (cm³)	Volume added (cm³)
rough	0.0	15.4	15.4
1st	15.4	30.4	15.0
2nd	30.4	45.4	15.0

Figure 5.12 The student's titration results

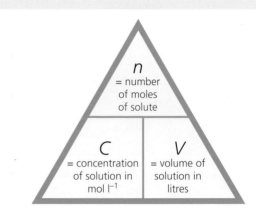

Figure 5.13 Formula triangle for a solution. Remember, this triangle gives:

- $n = C \times V$
- $C = n/V$
- $V = n/C$

Using the formula triangle again, we can calculate our final answer:

$$\text{concentration of NaOH} = \frac{n}{V}$$

$$= \frac{0.003}{0.02}$$

$$= 0.15\,mol\,l^{-1}$$

Checklist for Revision

- I understand that water and solutions made using water as the solvent contain both hydrogen and hydroxide ions in varying concentrations.
- I understand that pH is related to the concentration of hydrogen ions in a solution.
- I understand that neutralisation is a reaction between an acid and a base which results in the pH moving to 7.
- I can calculate the concentration or volume of a solution of an acid or an alkali given the results of a titration.

End-of-chapter questions

1 Which of the following compounds could be produced in the reaction between an acid and an alkali?
   - A Hydrogen chloride
   - B Calcium hydroxide
   - C Potassium sulfate
   - D Nitrogen dioxide

2 Which of the following compounds would dissolve in water to produce a solution with a pH of less than 7?
   - A MgO
   - B $CaCO_3$
   - C Na_2O
   - D CO_2

3 What is the concentration of a solution containing 0.8 mol of potassium chloride in $200\,cm^3$ of solution?
   - A $0.004\,mol\,l^{-1}$
   - B $0.4\,mol\,l^{-1}$
   - C $4\,mol\,l^{-1}$
   - D $40\,mol\,l^{-1}$

4 Identify the spectator ions in the following equation:
$$2HCl + Mg(OH)_2 \longrightarrow MgCl_2 + 2H_2O$$
   - A H^+ and Mg^{2+}
   - B H^+ and OH^-
   - C Mg^{2+} and Cl^-
   - D Mg^{2+} and OH^-

5 Complete the following word equations:
   a) nitric acid + potassium oxide →
   b) sulfuric acid + lithium carbonate →
   c) sodium hydroxide + nitric acid →
   d) hydrochloric acid + ammonium hydroxide →

6 A student titrated $25\,cm^3$ portions of $0.1\,mol\,l^{-1}$ nitric acid solution against $0.1\,mol\,l^{-1}$ sodium hydroxide. Her results are shown in Table 5.7.

Titration	Volume of sodium hydroxide (cm^3)
1	26.8
2	24.9
3	24.7

Table 5.7

   a) Name a piece of apparatus she could use to measure $25\,cm^3$ portions of the nitric acid accurately.
   b) What type of reaction is taking place between the nitric acid and the sodium hydroxide?

   c) How many moles of nitric acid reacted?
   d) What is the average volume of sodium hydroxide required to neutralise $25\,cm^3$ of the acid?

7 A student decided to produce potassium sulfate by first dissolving 276 g of potassium carbonate in water and then making the volume of the solution up to exactly 1 litre by adding more water. He added potassium carbonate solution to sulfuric acid to produce the desired salt.
   a) Write a word equation for the reaction.
   b) Rewrite this equation using chemical formulae.
   c) What is the concentration of the potassium carbonate solution?

8 Explain the meaning of the following terms:
   a) a base
   b) an alkali.

9 A student wanted to find out the concentration of a solution of lithium hydroxide. She completed a titration experiment and found that $10.0\,cm^3$ of lithium hydroxide was neutralised by an average volume of $12.0\,cm^3$ of $0.1\,mol\,l^{-1}$ hydrochloric acid solution. Calculate the concentration of her lithium hydroxide solution.
$$LiOH(aq) + HCl(aq) \longrightarrow LiCl(aq) + H_2O(l)$$

10 $15.0\,cm^3$ of a solution of $0.1\,mol\,l^{-1}$ sodium hydroxide neutralises $60.0\,cm^3$ of a solution of hydrochloric acid. What is the concentration of the acid?
$$HCl(aq) + NaOH(aq) \longrightarrow NaCl(aq) + H_2O(l)$$

11 $25.0\,cm^3$ of a solution of $0.2\,mol\,l^{-1}$ potassium hydroxide neutralises $35.0\,cm^3$ of a solution of nitric acid. What is the concentration of the acid?
$$HNO_3(aq) + KOH(aq) \longrightarrow KNO_3(aq) + H_2O(l)$$

12 What volume of a solution of $0.1\,mol\,l^{-1}$ sodium hydroxide is needed to neutralise $20.0\,cm^3$ of a solution of $0.3\,mol\,l^{-1}$ hydrochloric acid?
$$HCl(aq) + NaOH(aq) \longrightarrow NaCl(aq) + H_2O(l)$$

13 What volume of a solution of $0.25\,mol\,l^{-1}$ sulfuric acid is needed to neutralise $40.0\,cm^3$ of a solution of $0.2\,mol\,l^{-1}$ lithium hydroxide?
$$H_2SO_4(aq) + 2LiOH(aq) \longrightarrow Li_2SO_4(aq) + 2H_2O(l)$$

14 $25.0\,cm^3$ of a solution of $0.2\,mol\,l^{-1}$ barium hydroxide neutralises $40.0\,cm^3$ of a solution of hydrochloric acid. What is the concentration of the acid?
$$2HCl(aq) + Ba(OH)_2(aq) \longrightarrow BaCl_2(aq) + 2H_2O(l)$$

15 $25.0\,cm^3$ of a solution of $0.1\,mol\,l^{-1}$ sodium hydroxide neutralises $60.0\,cm^3$ of a solution of hydrochloric acid. What is the concentration of the acid?

$$HCl(aq) + NaOH(aq) \longrightarrow NaCl(aq) + H_2O(l)$$

16 What volume of a solution of $0.01\,mol\,l^{-1}$ potassium hydroxide solution is needed to neutralise $10.0\,cm^3$ of a solution of $0.15\,mol\,l^{-1}$ sulfuric acid?

$$H_2SO_4(aq) + 2KOH(aq) \longrightarrow K_2SO_4(aq) + 2H_2O(l)$$

17 What volume of a solution of $0.005\,mol\,l^{-1}$ nitric acid is needed to neutralise $15.0\,cm^3$ of a solution of $0.1\,mol\,l^{-1}$ lithium hydroxide?

$$HNO_3(aq) + LiOH(aq) \longrightarrow LiNO_3(aq) + H_2O(l)$$

This final question is very difficult – a real test of your knowledge!

18 $75.0\,cm^3$ of a solution of $0.1\,mol\,l^{-1}$ NaOH neutralises $25.0\,cm^3$ of a solution of an acid. The formula of the acid is H_xA and the concentration of the acid is $0.1\,mol\,l^{-1}$. What is the value of x?

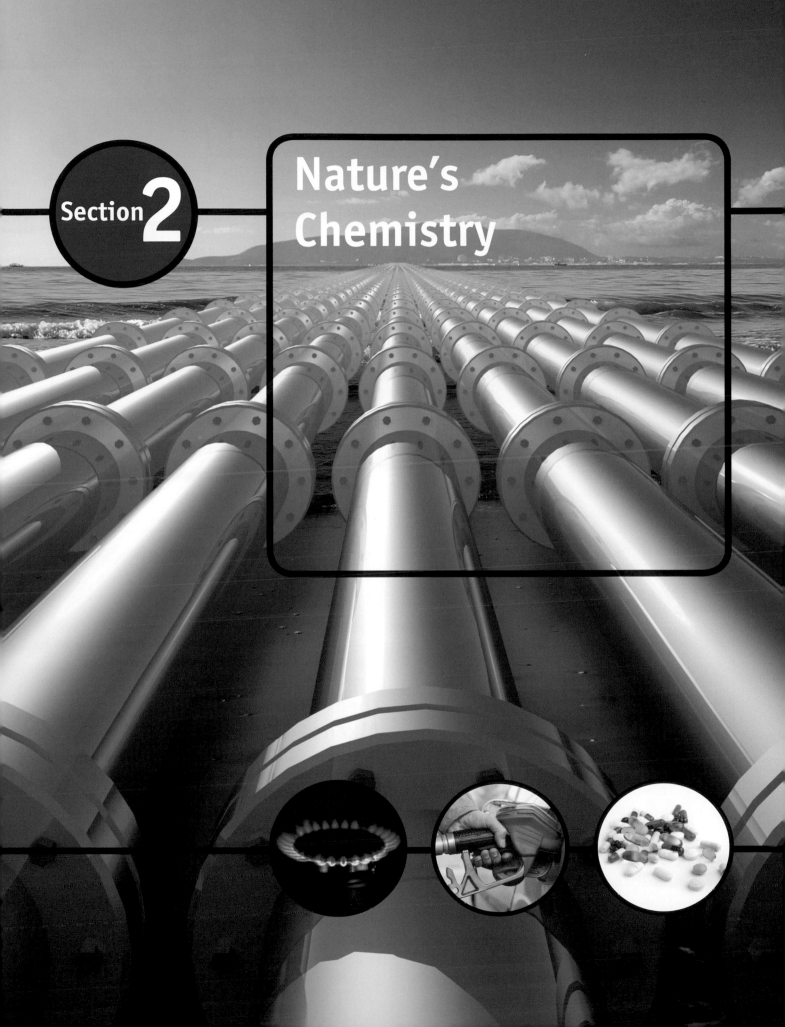

Section 2

Nature's Chemistry

6 Homologous series

It is hard to imagine the world without carbon compounds. From the fuels that power cars (Figure 6.1) to the deodorants and perfumes (Figure 6.2) that keep us smelling pleasant, carbon compounds are all around us.

Figure 6.2 Many perfumes contain fruity smelling carbon compounds

To make our study of carbon compounds much easier, chemists have collected similar compounds into families. A family of compounds is known as a **homologous series**.

Alkanes

Hydrocarbons are compounds that contain hydrogen and carbon only. The **alkanes** are a homologous series of hydrocarbons where all the carbon-to-carbon bonds are *single* covalent bonds. The first member of the alkane homologous series is methane, the fuel in natural gas. Figure 6.3 gives a three-dimensional model of methane, showing its tetrahedral shape. The three-dimensional shape of methane is easier to picture if you build a molecule of methane from molecular models. This tetrahedral shape is difficult to draw by hand, so it is often more convenient to show methane by drawing its structural formula, as shown in Figure 6.4, or by writing its molecular formula, CH_4.

Figure 6.1 Petrol and diesel contain carbon compounds such as alkanes and cycloalkanes

In the next two chapters, we will explore the structures and properties of three types of carbon compounds: hydrocarbons, alcohols and carboxylic acids.

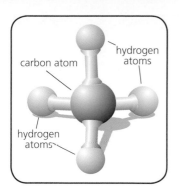

Figure 6.3 A molecule of methane

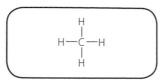

Figure 6.4 The full structural formula of methane

For large alkanes, such as hexane, we can also write a shortened structural formula. This is shown in Figure 6.5.

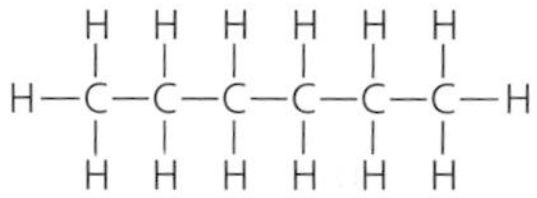

$$\begin{array}{cccccc} H & H & H & H & H & H \\ | & | & | & | & | & | \\ H-C-C-C-C-C-C-H \\ | & | & | & | & | & | \\ H & H & H & H & H & H \end{array}$$

$CH_3CH_2CH_2CH_2CH_2CH_3$ or $CH_3(CH_2)_4CH_3$

Figure 6.5 The full and shortened structural formulae of hexane

Table 6.1 lists the first eight members of the alkanes, showing their names and formulae.

Number of C atoms	Name	Molecular formula	Full structural formula	Shortened structural formula	State at room temperature (25 °C)
1	methane	CH_4	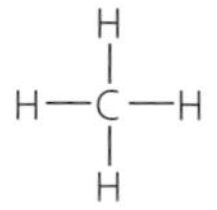	CH_4	gas
2	ethane	C_2H_6	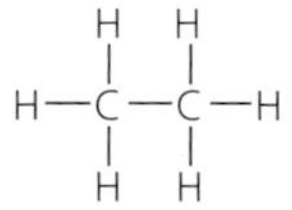	CH_3CH_3	gas
3	propane	C_3H_8	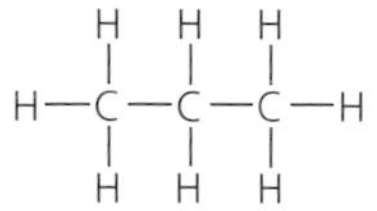	$CH_3CH_2CH_3$	gas
4	butane	C_4H_{10}	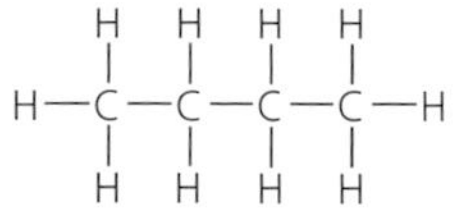	$CH_3CH_2CH_2CH_3$	gas
5	pentane	C_5H_{12}	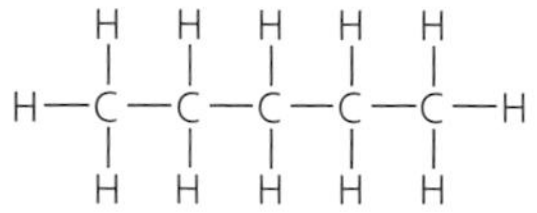	$CH_3CH_2CH_2CH_2CH_3$	liquid
6	hexane	C_6H_{14}	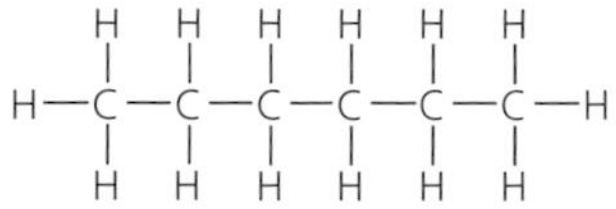	$CH_3CH_2CH_2CH_2CH_2CH_3$	liquid
7	heptane	C_7H_{16}	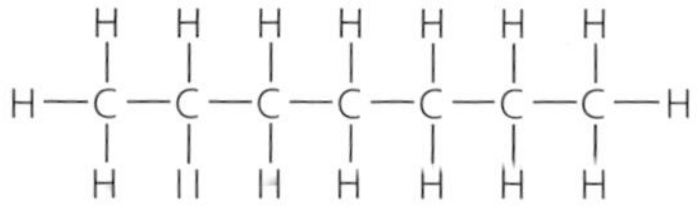	$CH_3CH_2CH_2CH_2CH_2CH_2CH_3$	liquid
8	octane	C_8H_{18}	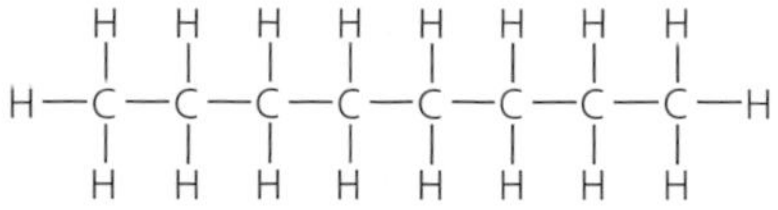	$CH_3CH_2CH_2CH_2CH_2CH_2CH_2CH_3$	liquid

Table 6.1 The alkanes

General formula

For any homologous series we can write a general formula. This allows us to write the molecular formula for any member of a series where the number of carbon atoms is known. For the alkanes, the general formula is: C_nH_{2n+2}. For example, for an alkane with four carbon atoms ($n = 4$), the number of hydrogen atoms would be $(2 \times 4) + 2 = 10$. The formula would be C_4H_{10}, which is the correct formula for butane.

Physical properties

A physical property is something that can be observed or measured. Examples include melting and boiling points, viscosity and colour. If you look up the melting and boiling point data for the alkanes, you will notice that there is a gradual increase from methane to octane. Bigger alkane molecules are more strongly attracted to each other than smaller alkane molecules and so it takes more energy to separate the bigger alkane molecules from each other. Hence, boiling points increase as the alkane molecules increase in size. Another property of the alkanes worth noting is that they are insoluble in water.

Chemical properties and uses

Chemical properties of substances refer to the chemical reactions which they take part in. The most common chemical reaction of the alkanes is combustion. You will be familiar with burning methane from experiments using the Bunsen burner (Figure 6.6) since methane is the fuel in natural gas. When alkanes burn completely, they produce carbon dioxide and water:

$$CH_4 + O_2 \longrightarrow CO_2 + H_2O$$

Complete combustion occurs in the Bunsen burner when the air hole is fully open. However, you will be aware that if you close the air hole, the flame colour changes from blue to yellow (Figure 6.7). Closing the air hole reduces the flow of oxygen. This causes incomplete combustion to occur, which accounts for the fact that the flame becomes sooty since carbon is one of the products of incomplete combustion. The toxic gas carbon monoxide is also produced during incomplete combustion.

The fact that alkanes burn makes them very useful as fuels. Table 6.2 lists some of the common uses for some of the alkanes.

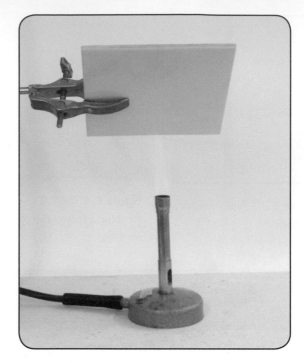

Figure 6.6 Complete combustion of methane produces carbon dioxide and water

Figure 6.7 Carbon monoxide and carbon (soot) are also produced when methane burns in a poor supply of oxygen

Name of alkane	Use
methane	the fuel in natural gas for cooking and heating (Figure 6.8)
butane	the fuel used in barbecues and camping stoves (Figure 6.9)
octane	one of the compounds in petrol (Figure 6.10)

Table 6.2 Uses for some of the common alkanes

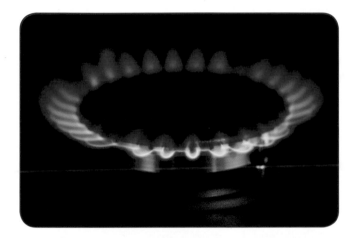

Figure 6.8 Methane is used at home for cooking on gas hobs

Figure 6.9 Butane or a butane/propane mix is used in camping stoves and can be bought in cylinders

Figure 6.10 Octane is one of the compounds found in petrol

Homologous series – a definition

Our study of the alkanes has revealed that the members of the alkane family share some common features:

- They have the same general formula, C_nH_{2n+2}.
- They have similar chemical properties; for example, they all burn.
- They show a gradual change in physical properties; for example, boiling points increase as the alkane molecules increase in size.

These features are key to being part of a homologous series. In other words, a group of compounds that have the same general formula, similar chemical properties and show a gradual change in physical properties can be said to belong to the same homologous series.

Isomers

If you use a set of molecular models to build the structure of a compound with molecular formula C_2H_6, there is only one possible structure. This is shown in Figure 6.11.

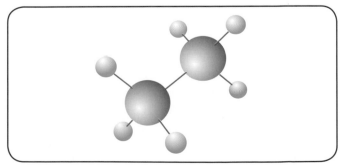

Figure 6.11 Molecular model structure of C_2H_6

However, if you try to build a molecular model structure of a compound with formula C_4H_{10}, you find that there are two possible structures, as shown in Figure 6.12.

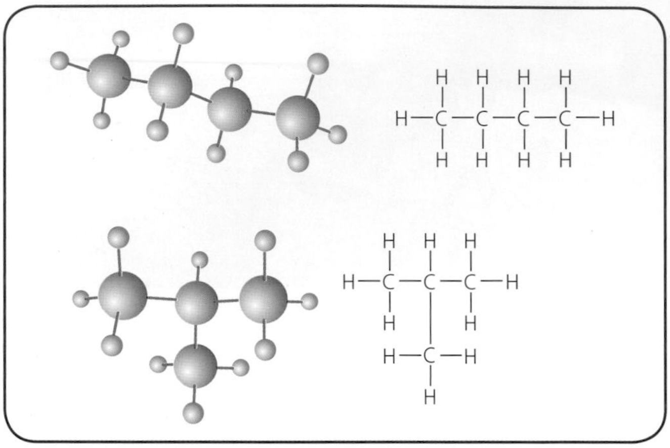

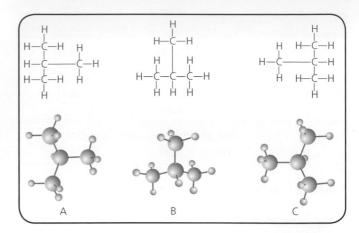

Figure 6.13 Rotating a structure – it is still the same isomer. Rotate molecule A 90° anticlockwise to obtain molecule B. Rotate molecule B 90° anticlockwise to obtain molecule C

Figure 6.12 Structures with the formula C_4H_{10}

Compounds that have the same molecular formula but different structural formulae are known as **isomers**. Isomers have different physical properties. For example, the boiling points of the two isomers of butane shown in Figure 6.12 are different. As the alkanes become larger in size, the number of isomers for each alkane increases too. For example, pentane has three isomers and hexane has five isomers.

Building molecular models is a useful way of determining how many isomers a compound has. For example, the second isomer of butane shown in Figure 6.12 can be drawn in many different ways, as shown in Figure 6.13, but they all show the exact same isomer! If you build the molecular model for this isomer, you will see that the structures simply show different rotations of the molecule. Try this to find out for yourself.

Systematic naming

If you look at the label of ingredients from a bottle of deodorant (Figure 6.14) or shower gel, you will notice that there are many exotic names.

In National 5 Chemistry you will learn how to identify and draw the structures of some of these compounds using **systematic naming**. This is particularly useful for identifying isomers of a compound as systematic naming gives isomers different names. For example, in

Figure 6.14 Deodorants contain carbon compounds such as alcohols

Figure 6.12 only the top structure (the molecule with its four carbon atoms in a straight chain) is called butane. The lower branched molecule is called methylpropane.

Here are some simple rules for naming branched alkanes:

1 Identify the longest chain of carbon atoms and name the alkane with this number of carbon atoms.
2 Identify the branch and name it according to the number of carbon atoms in the branch.
3 Number the branch so that it has the lower of two possible numbers.

Worked examples

1 Let us apply these rules to the isomer of pentane shown in Figure 6.15.

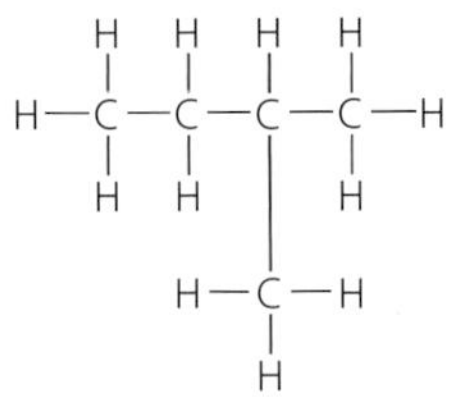

Figure 6.15 Naming an isomer of pentane

1 The longest chain of carbon atoms is four, so the alkane name is butane.

2 There is one branch. Branches are named after the alkane, as shown in Table 6.3. In this case, the branch is a methyl branch. If there had been two methyl branches, the prefix 'di' would have been used. In other words, it would be called 'dimethyl'.

Name of branch	Structure	Number of branches	Prefix
methyl	$-CH_3$	1	–
ethyl	$-C_2H_5$	2	di
propyl	$-C_3H_7$	3	tri

Table 6.3 Naming branches

3 In this example, the only place a methyl group can go on a 4-carbon chain is on the second carbon atom. Hence the structure is called methylbutane. We do not need to use the number '2' (as in 2-methylbutane) since there is only one way to draw methylbutane: with the methyl group on the second carbon atom of the 4-carbon chain. Look at Figure 6.16. It shows two seemingly different ways of drawing a 4-carbon chain with a methyl group. However, in both cases, the methyl group is on the second carbon atom of the chain. Hence, both structures are the same; they are both methylbutane.

If you are not convinced, try building molecule A and then rotating it to look like molecule B.

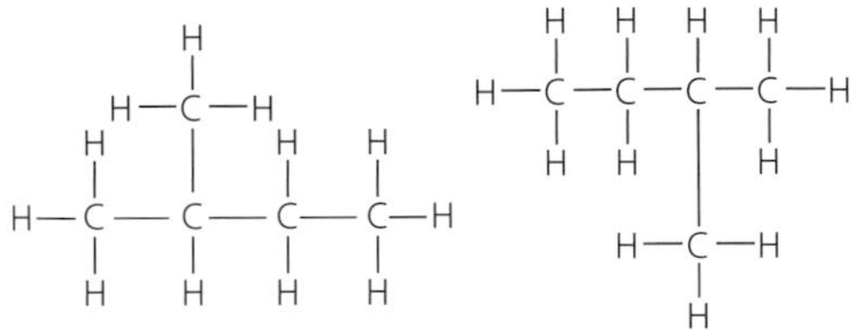

Figure 6.16 Representations of methylbutane

2 We will now apply the same rules to the structure given in Figure 6.17.

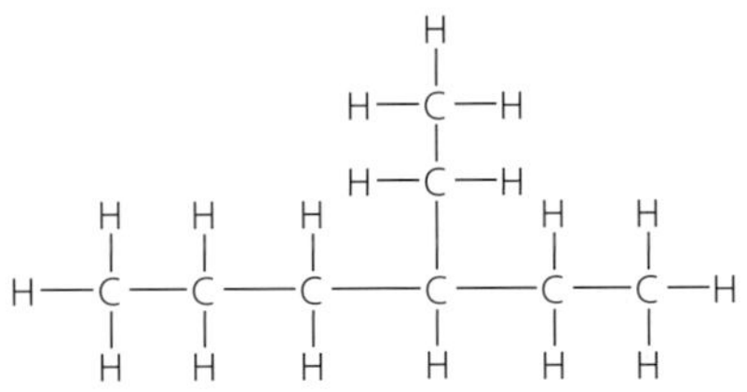

Figure 6.17

1 The longest chain has six carbon atoms: hexane.
2 There is one branch: ethyl.
3 The branch is on position 3.

This compound is called: 3-ethylhexane.

3 Applying the same rules to the structure given in Figure 6.18, we can work out that the compound is called 3,4-dimethylhexane.

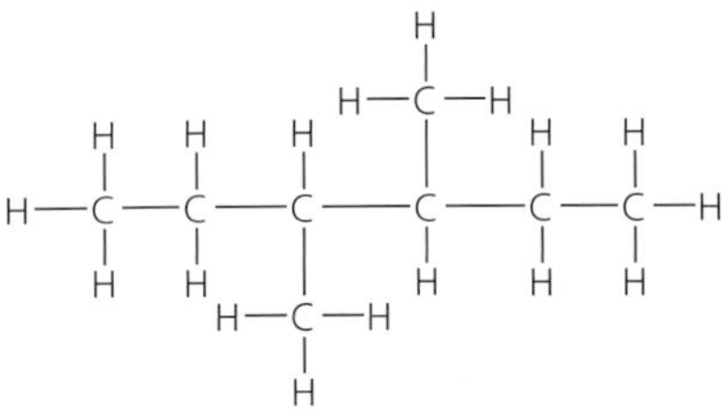

Figure 6.18

Questions

1 Draw full structural formulae for the following:
a) 3-methylpentane
b) 3,3-dimethylpentane
c) 2,2,3-trimethylpentane.

Cycloalkanes

So far our study of hydrocarbons has dealt with the alkanes where each carbon is bonded to a neighbouring carbon atom in a straight chain. The **cycloalkanes** are a homologous series of hydrocarbons where the carbon atoms join together to form a closed ring. The first member of the cycloalkane series is cyclopropane, shown in Figure 6.19.

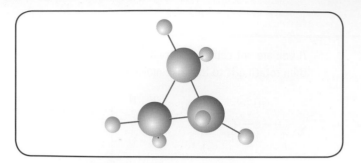

Figure 6.19 Cyclopropane, the first member of the cycloalkanes

You will see from Table 6.4 that the cycloalkanes are named after the parent alkane. For example, butane has four carbon atoms, so the cycloalkane with four carbon atoms is called cyclobutane.

General formula

The general formula for a cycloalkane is C_nH_{2n}.

Physical properties

As the cycloalkanes increase in size, their melting and boiling points increase. Like the alkanes, the cycloalkanes are insoluble in water.

Number of C atoms	Name	Molecular formula	Full structural formula	Shortened structural formula	State at room temperature (25 °C)
3	cyclopropane	C_3H_6			gas
4	cyclobutane	C_4H_8			gas
5	cyclopentane	C_5H_{10}			liquid
6	cyclohexane	C_6H_{12}			liquid
7	cycloheptane	C_7H_{14}			liquid
8	cyclooctane	C_8H_{16}			liquid

Table 6.4 The cycloalkanes

Chemical properties

Cycloalkanes burn to produce carbon dioxide and water.

Uses

Cycloalkanes can be used as fuels, but most are used to make other compounds. For example, the main use of cyclohexane is for making nylon. Like the alkanes, cycloalkanes do not dissolve in water so they are good solvents for other compounds that do not dissolve in water.

The alkenes

The **alkenes** are a homologous series of hydrocarbons that contain at least one carbon-to-carbon double bond. Table 6.5 gives the names and formulae for the first seven alkenes.

General formula

The general formula for the alkenes is C_nH_{2n}.

Isomers and systematic naming

The fact that the cycloalkanes and the alkenes have the same general formula, C_nH_{2n}, allows us to conclude that cycloalkanes are isomers of the corresponding alkene with the same number of carbon atoms. For example, the cycloalkane with three carbon atoms is cyclopropane, C_3H_6. This is an isomer of propene, also C_3H_6. Both compounds have the same molecular formula but a different structural formula, as shown in Figure 6.20.

Number of C atoms	Name	Molecular formula	Full structural formula	Shortened structural formula	State at room temperature (25 °C)
2	ethene	C_2H_4	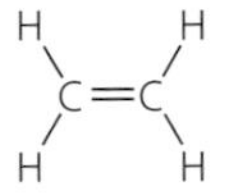	$CH_2{=}CH_2$	gas
3	propene	C_3H_6	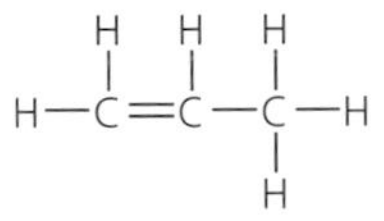	$CH_2{=}CHCH_3$	gas
4	butene	C_4H_8	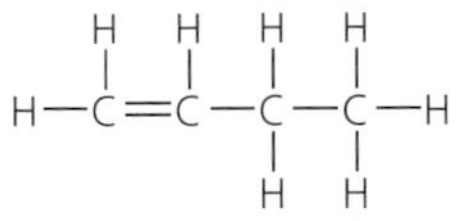	$CH_2{=}CHCH_2CH_3$	gas
5	pentene	C_5H_{10}	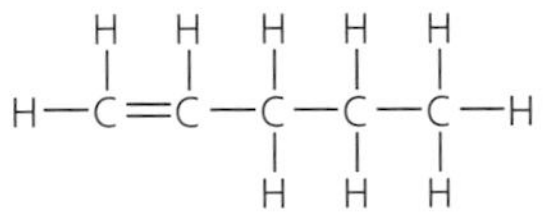	$CH_2{=}CHCH_2CH_2CH_3$	liquid
6	hexene	C_6H_{12}	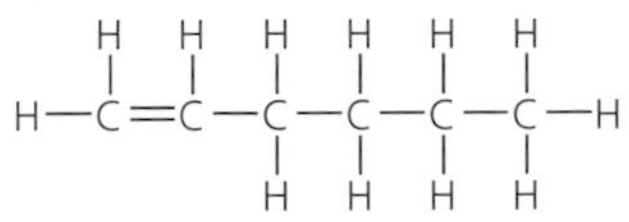	$CH_2{=}CHCH_2CH_2CH_2CH_3$	liquid
7	heptene	C_7H_{14}		$CH_2{=}CHCH_2CH_2CH_2CH_3$	liquid
8	octene	C_8H_{16}	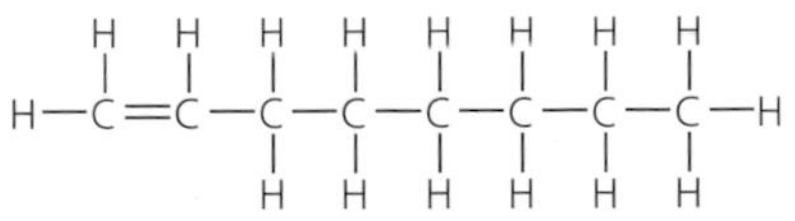	$CH_2{=}CHCH_2CH_2CH_2CH_2CH_3$	liquid

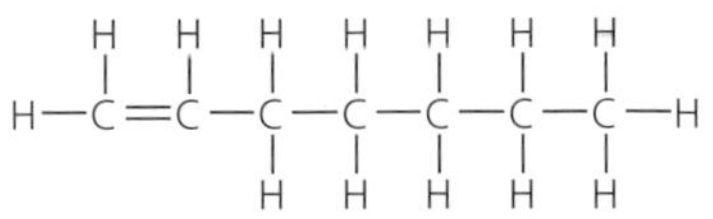

Table 6.5 The alkenes

55

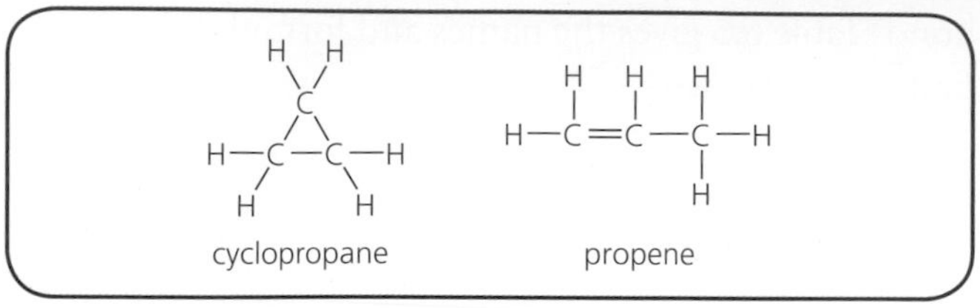

Figure 6.20 Two isomers with formula C_3H_6

Alkenes can also form isomers by moving the position of the double bond. For example, two isomers of butene can be formed by moving the double bond, as shown in Figure 6.21.

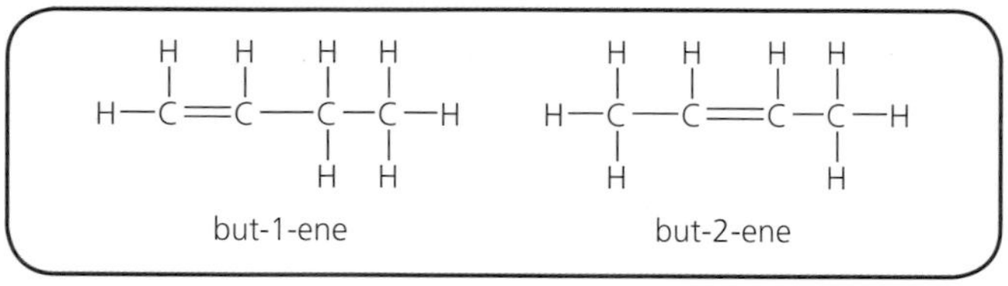

Figure 6.21 Two alkene isomers of butene

The isomers are named by numbering the carbon atoms from the end that gives the carbon of the double bond the lower number. For example, the isomer of pentene shown in Figure 6.22 would be called pent-2-ene rather than pent-3-ene.

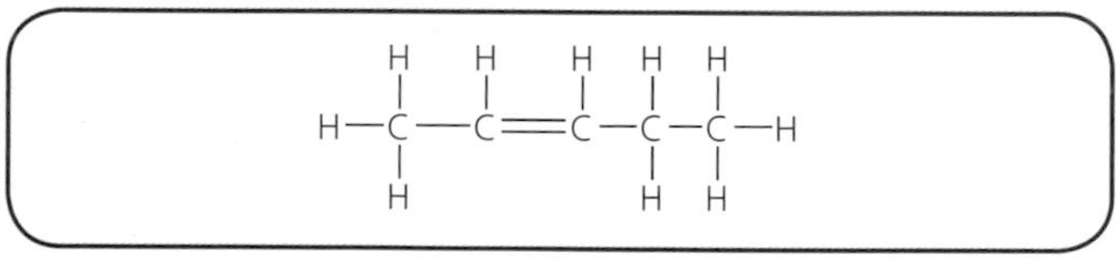

Figure 6.22 Naming an isomer of pentene

Where there are branches, the double bond takes priority over the branch. This is shown in the examples in Table 6.6.

Compound structure	Name
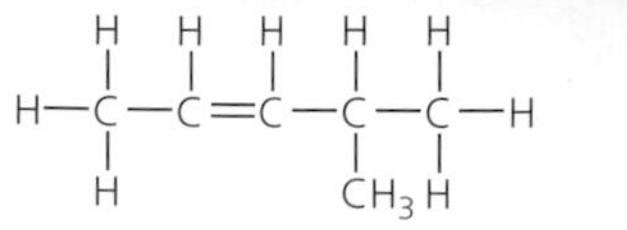	4-methylpent-2-ene
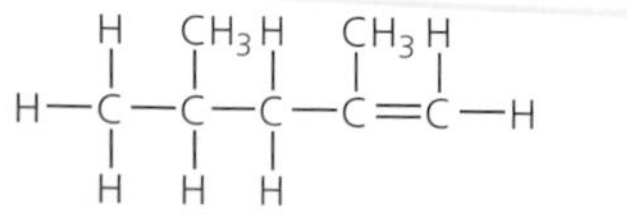	2,4-dimethylpent-1-ene
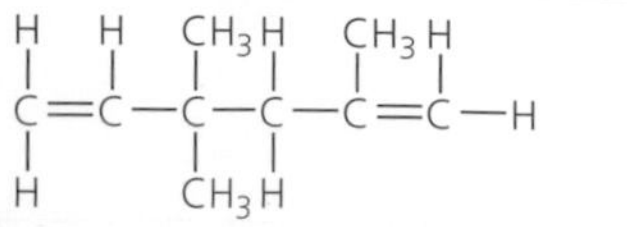	2,4,4-trimethylhex-1,5-diene

Table 6.6

The third structure in Table 6.6 does not conform to the general formula for the alkenes, C_nH_{2n}. It is an example of an unsaturated hydrocarbon that contains more than one double bond. When this occurs, a prefix is used to indicate the number of double bonds: diene for two double bonds, triene for three double bonds, etc.

Physical properties

As the alkenes increase in size, their melting and boiling points increase. Like the alkanes and cycloalkanes, the alkenes are insoluble in water.

Chemical properties

Like other hydrocarbons, alkenes burn to produce carbon dioxide and water. Unlike our previous two families, in which the carbon atoms were all bonded to one another by single covalent bonds, alkenes contain a carbon-to-carbon double bond. They are said to be **unsaturated**. The presence of the carbon-to-carbon double bond allows alkenes to take part in reactions known as **addition reactions**, where reacting molecules can 'add' across the reactive double bond, as shown in the examples which follow.

Adding hydrogen
Alkenes can react with hydrogen gas to produce alkanes (Figure 6.23).

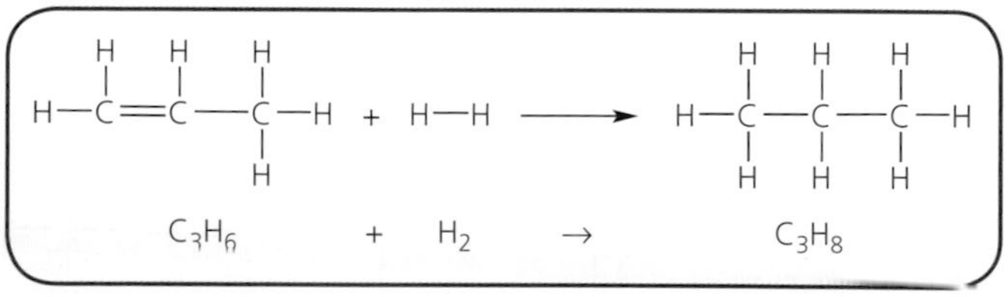

Figure 6.23 Reaction of propene with hydrogen gas

This reaction is known as **hydrogenation.** The product alkane is said to be **saturated** because no more atoms can be added to the molecule. The alkene, on the other hand, is an example of an unsaturated molecule as more atoms can bond to it when the double bond breaks.

Hydrogenation is used in food manufacturing to change the texture of food, for example to change liquid oils into solids. This involves reacting polyunsaturated oils, which have several double bonds, with hydrogen to change some or all of the double bonds into single bonds. If you look at a cake or biscuit wrapper, you will often see the phrase 'contains hydrogenated vegetable oil' on the label of ingredients (Figure 6.24). However, due to health concerns, many food manufacturers have stopped using hydrogenated oils, which you may also see mentioned on the label.

Figure 6.24 Hydrogenation is used in the food industry

Adding bromine

A convenient laboratory test for the presence of double bonds in a molecule is to shake the compound with bromine solution. Alkenes cause the bromine solution to decolourise rapidly (Figure 6.25). With alkanes and cycloalkanes, the bromine solution does not decolourise rapidly. The reaction of an alkene with bromine solution is another example of an addition reaction: the bromine molecule adds across the double bond of the alkene, as shown in Figure 6.26.

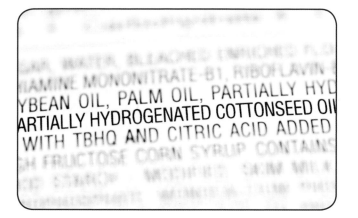

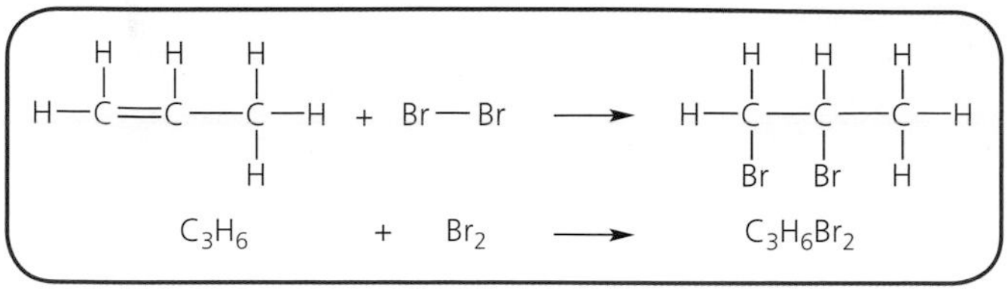

Figure 6.26 Reacting propene with bromine solution

Reaction with halogens

Alkenes can react with other halogens in the same manner as they react with bromine. Since the products always contain two halogen atoms, they are known as dihaloalkanes.

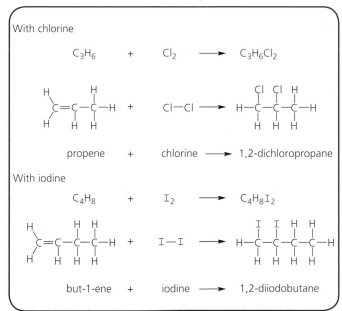

Figure 6.27 Reacting alkenes with halogens

Uses

Making plastics

Alkenes can be made to react together to form very large compounds known as **polymers** (Figure 6.28). For example, ethene molecules can react with other ethene molecules to form poly(ethene); propene molecules can react with other propene molecules to form poly(propene). You will find out more about this reaction, **addition polymerisation**, and the uses of polymers in Chapter 10.

Figure 6.25 Alkenes rapidly decolourise bromine solution

ethanol, it is produced by the catalytic hydration of ethene (Figure 6.29):

$$C_2H_4 + H_2O \longrightarrow C_2H_5OH$$

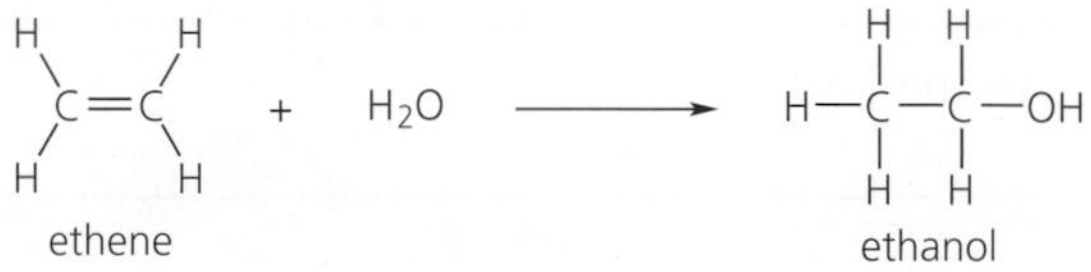

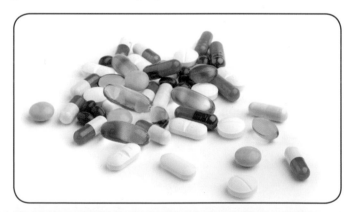

Figure 6.29 Ethene reacts with water to produce ethanol

This involves passing ethene gas and steam over an acidic catalyst at high temperature and pressure. The INEOS chemical company is the largest producer of synthetic ethanol (ethanol produced by the catalytic hydration of ethene) in the world. Its site in Grangemouth, Scotland, produces over 300 000 tonnes of ethanol every year by this process. This ethanol is then sold to other manufacturers where it is used extensively as a solvent, for making medicines (Figure 6.30), printing inks and cosmetic products such as hairspray.

Figure 6.28 Alkenes are reacted to form polymers

Industrial production of ethanol

In the next chapter, you will meet a new family of compounds known as the **alcohols**. The process of fermentation can be used to produce the alcohol ethanol from plants such as grapes, apples and barley. This is how the alcoholic drinks wine, cider and whisky are produced. Ethanol and other alcohols can also be produced by reacting alkenes with water. This addition reaction is known as catalytic **hydration**. For example, in order to meet the industrial demand for the alcohol

Figure 6.30 Ethanol is a useful solvent for making medicines and cosmetic products

Checklist for Revision

- I can name and write formulae for the alkanes, alkenes and cycloalkanes.
- I can use systematic naming to write a name for a structure.
- I can draw a structure for a compound from its systematic name.
- I know what is meant by the word 'isomer' and can identify isomers of a compound.
- I know what is meant by the term 'homologous series'.

- I know what is meant by the words 'saturated' and 'unsaturated'.
- I know some of the physical properties of alkanes, cycloalkanes and alkenes.
- I know that alkenes take part in addition reactions and I can draw the products formed when alkenes react with halogens, hydrogen or water.
- I can describe how bromine solution can be used to distinguish between saturated and unsaturated compounds.

End-of-chapter questions

1 The grid below lists some common hydrocarbons.

A	B	C
C_2H_4	C_4H_{10}	Hexane
D	E	F
2-methylhex-1-ene	Cyclobutane	Butene

a) Identify the two alkanes.

b) Identify the two saturated hydrocarbons with four carbon atoms.

c) Identify the isomer of the compound in box F.

d) Identify the three compounds that would rapidly decolourise bromine solution.

e) Identify the compound with seven carbon atoms.

2 Use the grid below to answer the following questions.

A	B	C
	$CH_3CH_2CH_3$	
D	E	F
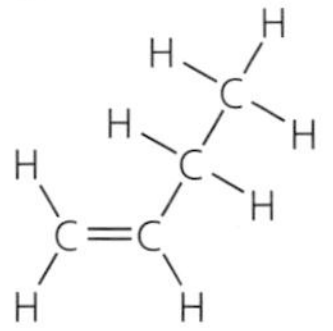	$CH_3CH_2CH_2CH_3$	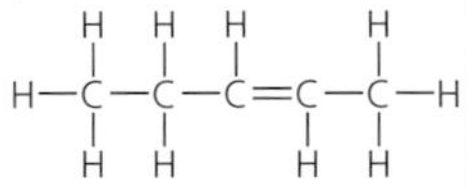

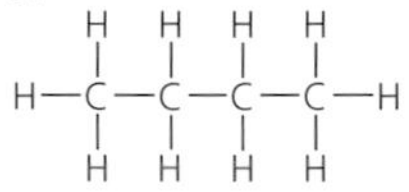

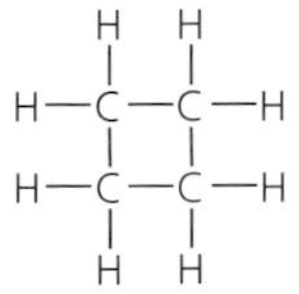

a) Identify the two boxes that could represent butane.

b) Identify the full structural formula for the compound formed when D reacts with hydrogen.

c) Identify the fourth member of the alkenes.

d) Identify the saturated isomer with formula C_4H_8.

e) Identify the two compounds that could take part in addition reactions.

3 Draw full structural formulae for the second member of

a) the alkanes

b) the alkenes

c) the cycloalkanes.

4 Two compounds, A and B, with formula C_3H_6 were reacted with bromine solution. The bromine solution decolourised rapidly when shaken with compound A, but did not decolourise rapidly with compound B. Draw full structural formulae for compounds A and B.

5 Draw full structural formulae for the following compounds:

a) 2-methylbut-1-ene

b) 3-methylpentane.

6 Write the systematic names for the following compounds.

a)

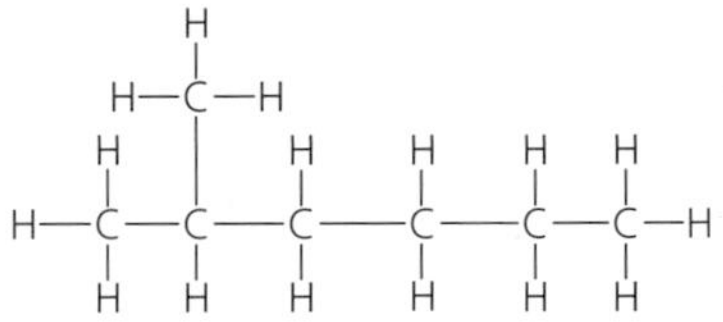

Figure 6.31

b)

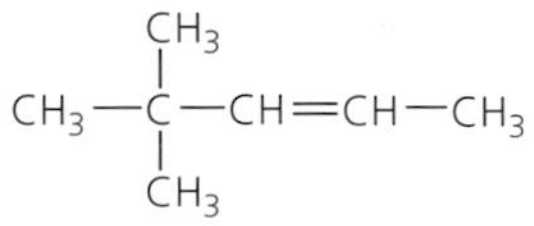

Figure 6.32

c)

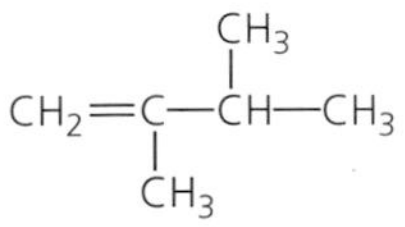

Figure 6.33

7 a) State the general formula for the alkanes.

b) Write molecular formulae for the alkanes with

i) 12 carbon atoms

ii) 80 hydrogen atoms.

8 The alkynes are a homologous series of hydrocarbons that contain a carbon-to-carbon triple bond. Table 6.7 lists the first three members of the alkynes.

Name	Structure
ethyne	H—C≡C—H
propyne	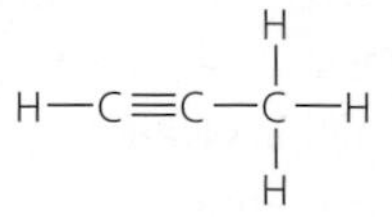
but-1-yne	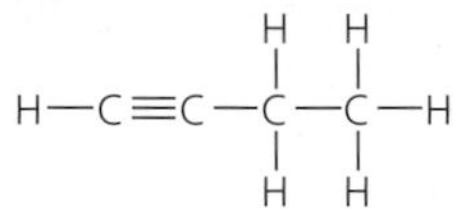

Table 6.7

a) Write the molecular formula for each alkyne.
b) The fourth member of the alkynes is pentyne. Draw a full structural formula for pent-1-yne.
c) Write a general formula for the alkynes.

9* Alcohols can be oxidised by hot copper(II) oxide. The product is either an aldehyde or a ketone (Figure 6.34).

Alcohol	Structural formula	Type of product	Structural formula
ethanol	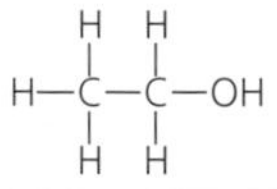	an aldehyde	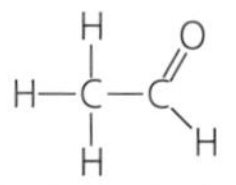
propan-1-ol	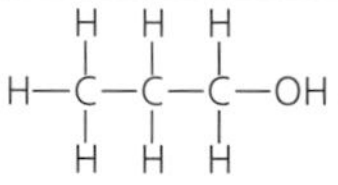	an aldehyde	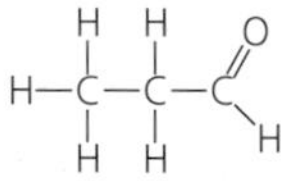
propan-2-ol	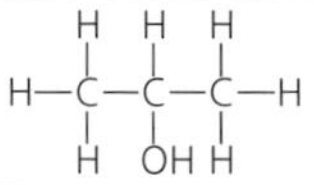	a ketone	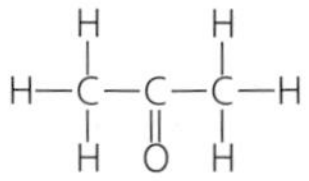
butan-2-ol	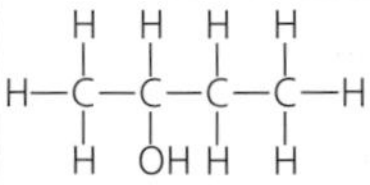	a ketone	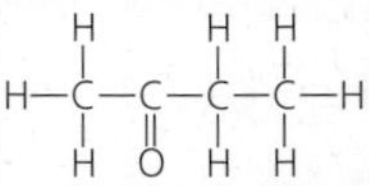

Figure 6.34

a) i) Aldehydes and ketones have the same general formula. Suggest a general formula for these compounds.
   ii) Write a general statement linking the type of product to the structure of the alcohol used.
b) Propan-1-ol can be formed by reacting an alkene with water. Draw the structure of the alkene used to produce propan-1-ol.
c) Draw a full structural formula for pentan-3-ol.

10* Dienes are a homologous series of hydrocarbons which contain two double bonds per molecule (Figure 6.35).

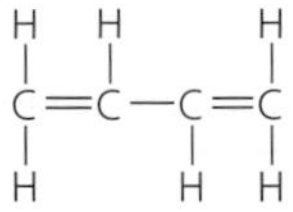

buta-1,3-diene

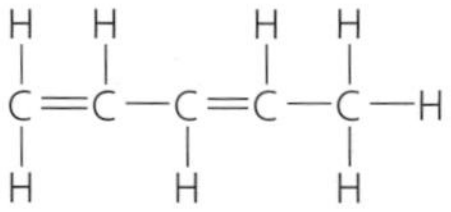

penta-1,3-diene

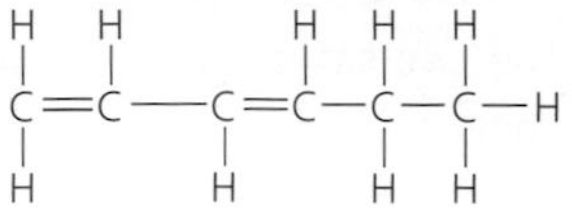

hexa-1,3-diene

Figure 6.35

a) What is meant by the term 'homologous series'?
b) Suggest a general formula for the dienes.
c) Write the molecular formula for the product of the complete reaction of penta-1,3-diene with bromine.
d) Draw a full structural formula for an isomer of buta-1,3-diene which contains only one double bond per molecule.

11 The following compounds were obtained by reacting alkenes with halogens.

A

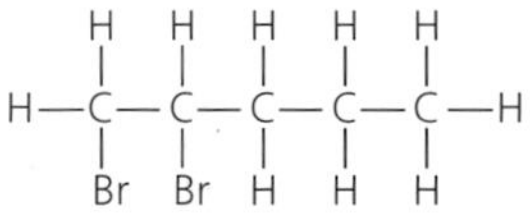

B

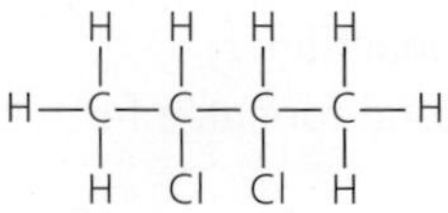

Figure 6.36

Draw structural formulae for the alkenes that could react with halogens to form:
a) molecule A
b) molecule B.

7 Everyday consumer products

Alcohols

Alcoholic drinks have been known of for thousands of years. The use and abuse of alcohol has been debated for much of that time and it can still be a great source of controversy. Many people use alcoholic drinks to celebrate social occasions (Figures 7.1 and 7.2) while other people do not drink alcohol. Unfortunately, most people in the UK are familiar with the tragic consequences of people consuming too much alcohol and the negative effects this can have on people and their families and friends. This drinking alcohol is known as ethanol and it is the first compound we will study in our review of the family of compounds known as the **alcohols**.

Figure 7.1 Ethanol is the alcohol in alcoholic drinks

Figure 7.2 People celebrating with alcohol

What is an alcohol?

The chemical structure of an alcohol can be obtained by replacing one of the hydrogen atoms of an alkane with an –OH group, the hydroxyl **functional group**, as shown in Figure 7.3.

Figure 7.3 Ethane and ethanol

As you can see from Table 7.1, replacing the final letter of the corresponding alkane by the name ending '-ol' gives the name of the alcohol.

General formula

The general formula for an alcohol is $C_nH_{2n+2}O$, although the shortened molecular formula $C_nH_{2n+1}OH$ is also used.

Name	Full structural formula	Shortened structural formula	Molecular formula
methanol		CH_3OH	CH_4O
ethanol		CH_3CH_2OH	C_2H_6O
propanol		$CH_3CH_2CH_2OH$	C_3H_8O
butanol		$CH_3CH_2CH_2CH_2OH$	$C_4H_{10}O$

Table 7.1 Names and formulae of common alcohols

Isomers

For straight-chain alcohols, a simple way of creating isomers involves moving the –OH onto a different carbon atom. The alcohol is then named so that the – OH group has the lowest number. For example, the two structures shown for butanol would be named butan-1-ol (Figure 7.4) and butan-2-ol (Figure 7.5).

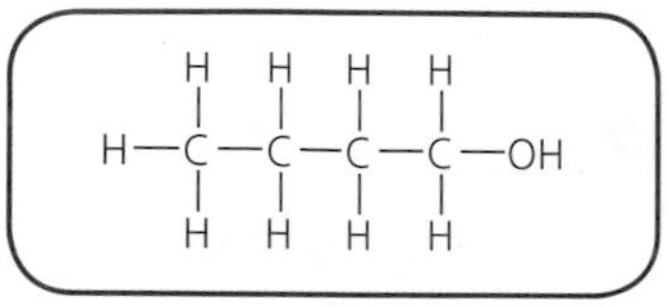

Figure 7.4 Butan-1-ol 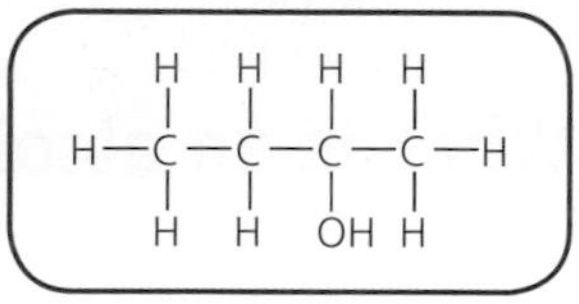   **Figure 7.5** Butan-2-ol

Questions

1 Draw full structural formulae for the following alcohols:
   a) hexan-2-ol
   b) pentan-3-ol.
2 Explain why hexan-4-ol would be an incorrect name for an isomer of hexanol.
3 State the names of the following alcohols:

   a)

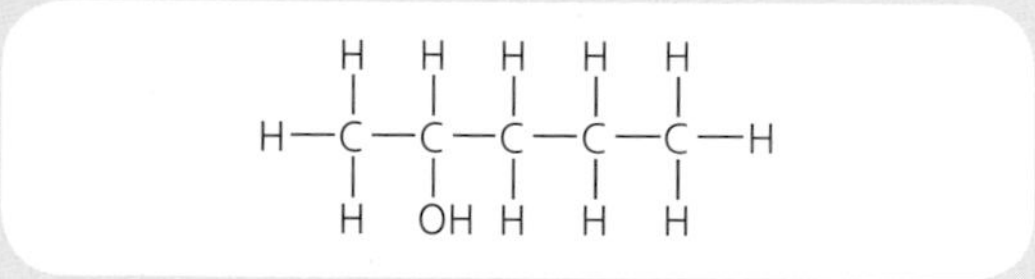

   Figure 7.6

   b)

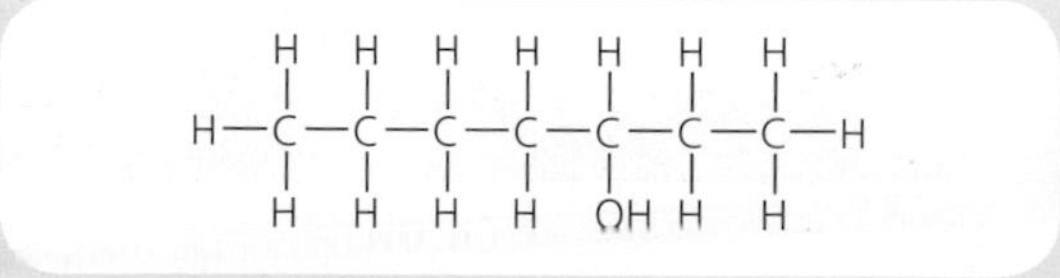

   Figure 7.7

Physical properties

Solubility of alcohols

Some people add water to whisky to enhance its flavour. The ethanol in whisky is soluble in water. The ability of alcohols to dissolve in water is affected by the length of their carbon chain. The short-chain alcohols (methanol, ethanol and propanol) all mix completely (are miscible) with water. Thereafter, the solubility decreases. For example, pentanol is much less soluble in water than ethanol.

Melting and boiling points

Part of the whisky making process relies on the fact that ethanol has a lower boiling point (78 °C) than water (100 °C), allowing them to be separated by distillation. Like the hydrocarbons, as the alcohols increase in size, their melting and boiling points increase. This is due to the increasing strength of intermolecular forces.

Uses of alcohols

Alcohols as solvents

One of the reasons we study alcohols in chemistry is because they are such useful compounds. The short-chain alcohols (methanol, ethanol, propanol and butanol) are fantastic solvents as they can dissolve a variety of compounds. They are frequently used to dissolve compounds to make perfumes, deodorants, paints and dyes. As the alcohols evaporate easily, they are ideal liquids for the uses described. Once they have been applied to the skin, a whiteboard or a wall, the alcohol will evaporate to leave the solute. The short-chain alcohols are also frequently used in cleaning where their ability to dissolve a variety of compounds and evaporate easily makes them an ideal choice. For example, screen wipes, baby wipes, alcohol gels (Figure 7.8) and disinfectant wipes often contain propan-2-ol and ethanol.

Figure 7.8 Hand gels, deodorants and baby wipes contain alcohols

Alcohols as fuels

If you have ever witnessed an alcohol being set alight, you will be well aware of how flammable they are. A common trick used by magicians (and chemistry teachers!) is to set alight a banknote (Figure 7.9). Flames emerge from the banknote but it does not burn. This works because the banknote has been soaked in ethanol and salt water. The water keeps the note cool while the ethanol burns. As long as the note stays cool, it will not burn. One of the problems with this demonstration is that the ethanol burns with a beautiful blue flame. In order to make the flames look more like 'fire', sodium chloride is often added since it burns with a yellow flame.

Figure 7.9 Setting alight a £10 note

This demonstration gives us two important facts about burning alcohols:

- They burn easily.
- They burn with a clean blue flame.

The flame demonstrates that alcohols burn 'cleanly'. In other words, the alcohol reacts with oxygen to produce carbon dioxide and water in a complete combustion reaction. For example:

$$CH_3OH + 1\tfrac{1}{2}O_2 \longrightarrow CO_2 + 2H_2O$$

As alcohols burn to release a significant amount of energy, they are commonly used as fuels. For example, methylated spirit (Figure 7.10) is a fuel that contains ethanol and methanol. Methylated spirit camping stoves are commonly used because the fuel is cheap and readily available, and as they can burn most liquid fuels, it makes them an attractive purchase for explorers

Figure 7.10 Methylated spirit contains ethanol and methanol and is commonly used as a fuel

trekking through remote parts of the world. To stop people drinking the fuel, various additives are used including a dye and compounds that give the liquid a foul taste.

Ethanol as a fuel

One of the major disadvantages of fossil fuels, such as oil, coal and natural gas, is that they are not renewable. In contrast, ethanol is a renewable fuel since it can be produced by fermentation of sugar cane, which can be grown continuously. The ethanol produced can be used as a fuel or mixed with petrol or diesel to produce a fuel blend (Figure 7.11). For example, 'Gasohol' is a petrol containing between 10% and 20% ethanol.

The use of pure ethanol as a fuel has been encouraged in certain countries, such as Brazil, where it is economic to produce it by fermentation using surplus **sucrose** from sugar cane. Thus sugar cane can be regarded as a renewable source of engine fuel. Since the cane took in carbon dioxide during its growth, the combustion

Figure 7.11 Ethanol can be used as a fuel for transport

Figure 7.12 Methanol is used as a fuel in dragster racing cars

of the ethanol to produce the same amount of carbon dioxide is not increasing the overall concentration of carbon dioxide in the atmosphere.

$$C_2H_5OH + 3O_2 \longrightarrow 2CO_2 + 3H_2O$$

Ethanol is therefore neutral in its contribution to 'greenhouse' gases.

Methanol as a fuel

Methanol is used as a fuel in racing cars in the USA (such as dragsters (Figure 7.12) and monster trucks) and has been suggested as a fuel for general use in cars too.

Carboxylic acids

Like many of the compounds we study in chemistry, the **carboxylic acids** are interesting because we find them in our everyday lives. For example, bee venom and ant stings contain methanoic acid (Figure 7.13), the first member of the carboxylic acids, whereas vinegar (Figure 7.14) contains a dilute solution of ethanoic acid, the second member of the series.

As you may have noticed the carboxylic acids we will study are named after the parent alkane with the term '-oic' added to the end of the name. Carboxylic acids contain the carboxyl functional group, shown in Figure 7.15.

Figure 7.13 Methanoic acid is found in bee venom and ant stings

Figure 7.14 Vinegar is a solution of ethanoic acid

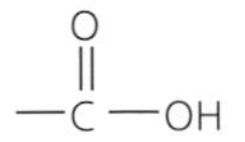

Figure 7.15 Carboxylic acids contain the carboxyl functional group

The names and structures of some carboxylic acids are listed in Table 7.2.

Name	Full structural formula	Shortened molecular formula	Molecular formula
methanoic acid	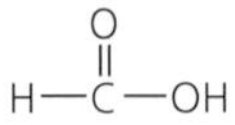	HCOOH	CH_2O_2
ethanoic acid	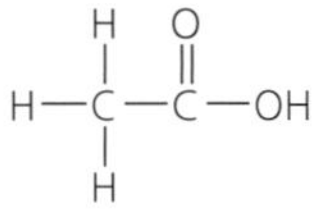	CH_3COOH	$C_2H_4O_2$
propanoic acid	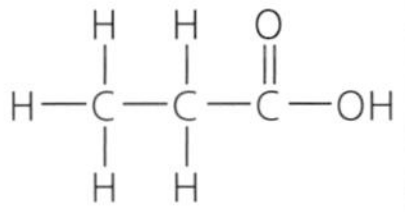	CH_3CH_2COOH	$C_3H_6O_2$

Table 7.2 Carboxylic acids

Carboxylic acids can be represented by the general formula $C_nH_{2n+1}COOH$. The structures shown in Table 7.2 show that a carboxylic acid is named like the alkane with the same number of carbon atoms. For example, the carboxylic acid with three carbon atoms takes the prefix 'prop-' from propane to become propanoic acid. Thus, the structure in Figure 7.16, with six carbon atoms, represents hexanoic acid.

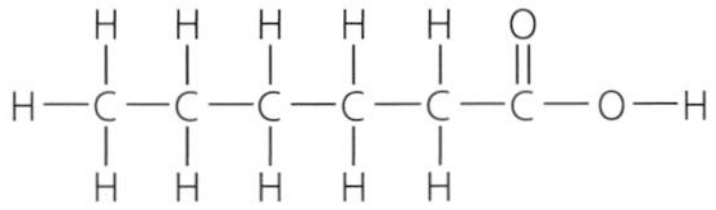

Figure 7.16 Hexanoic acid

Physical properties of carboxylic acids

Solubility
Like the alcohols, short-chain carboxylic acids (methanoic, ethanoic, propanoic and butanoic) are miscible with water. For example, vinegar is a solution of ethanoic acid. Thereafter, the solubility decreases as the size of the molecule increases.

Melting and boiling points
As carboxylic acids increase in size, their melting and boiling points increase due to the increasing strength of the intermolecular forces.

Questions

4 Draw full structural formulae for
   a) butanoic acid
   b) octanoic acid.
5 Name the following carboxylic acids:

a)

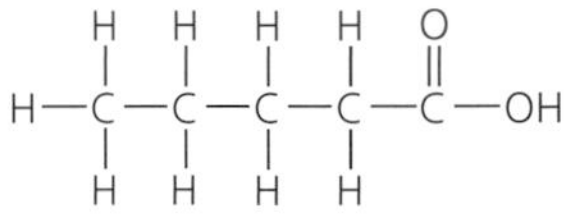

Figure 7.17

b)

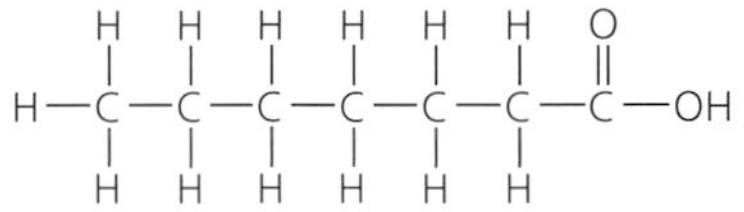

Figure 7.18

6 State the general formula for the carboxylic acids.

Carboxylic acids and acidity

As you discovered in Chapter 5, acids such as hydrochloric and sulfuric are acidic because they produce hydrogen ions (H^+) when they dissolve in water.

$$HCl(aq) \longrightarrow H^+(aq) + Cl^-(aq)$$

Carboxylic acids also produce H^+ ions when added to water:

$$CH_3COOH(aq) \rightleftharpoons CH_3COO^-(aq) + H^+(aq)$$

You also learned in Chapter 5 that the symbol $\rightleftharpoons$ is used to indicate a reversible reaction. Carboxylic acids only partially dissociate when added to water. The H^+ ions come from the carboxyl group. This is shown more clearly by drawing the full structural formula for ethanoic acid:

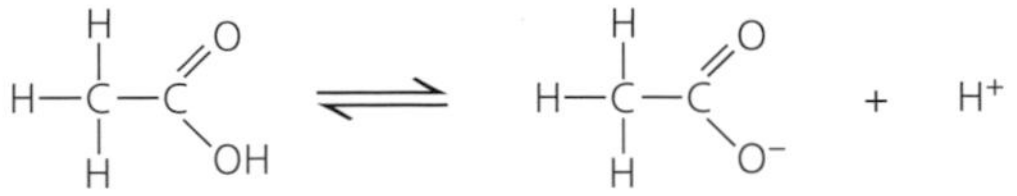

Figure 7.19 Dissociation of ethanoic acid

65

Carboxylic acids are known as **weak acids**. Ethanoic acid, like all water-soluble carboxylic acids, exists as a mixture of molecules and ions when in aqueous solution. Hydrochloric and sulfuric acid are known as **strong acids** since they break up completely into ions when they dissolve in water.

Reactions of carboxylic acids

Solutions of carboxylic acids have a pH less than 7 and, like other acids, they can take part in reactions with bases. For example, hydrochloric acid and nitric acid react with bases to form salts in neutralisation reactions:

$$HCl(aq) + NaOH(aq) \longrightarrow NaCl(aq) + H_2O(l)$$

$$2HNO_3(aq) + K_2O(aq) \longrightarrow 2KNO_3(aq) + H_2O(l)$$

The salt is formed by replacing the hydrogen ion (H^+) of the acid with the positive ion of the base (Na^+ and K^+ in the above examples).

Carboxylic acids can undergo the same types of reactions as the common laboratory acids; for example, they can be neutralised by:

- metal oxides to produce a salt and water
- metal hydroxides to produce a salt and water
- metal carbonates to produce a salt, water and carbon dioxide.

For example:

$$2CH_3COOH + Na_2O \longrightarrow 2CH_3COONa + H_2O$$

ethanoic acid   sodium oxide   sodium ethanoate   water

$$HCOOH + LiOH \longrightarrow HCOOLi + H_2O$$

methanoic acid   lithium hydroxide   lithium methanoate   water

$$2CH_3CH_2COOH + K_2CO_3 \longrightarrow 2CH_3CH_2COOK + H_2O + CO_2$$

propanoic acid   potassium carbonate   potassium propanoate   water   carbon dioxide

Carboxylic acids also react with metals to produce a salt and hydrogen gas.

Naming salts

For National 5 Chemistry, you should learn to name the salts formed from carboxylic acid reactions. This is achieved by:

i) Changing the ending of the carboxylic acid from '-oic' to '-oate'.

Carboxylic acid	Salt ending
methanoic	methanoate
ethanoic	ethanoate
propanoic	propanoate
butanoic	butanoate
pentanoic	pentanoate
hexanoic	hexanoate
heptanoic	heptanoate
octanoic	octanoate

Table 7.3

ii) Adding the positive ion name (usually the metal).

Carboxylic acid	Reacting with	Salt
methanoic	sodium hydroxide	sodium methanoate
ethanoic	potassium hydroxide	potassium ethanoate
propanoic	lithium oxide	lithium propanoate
butanoic	magnesium carbonate	magnesium butanoate
pentanoic	ammonium hydroxide	ammonium pentanoate
hexanoic	calcium oxide	calcium hexanoate
heptanoic	sodium carbonate	sodium heptanoate
octanoic	lithium hydroxide	lithium octanoate

Table 7.4

The following table shows the other product(s) formed when salts are made from carboxylic acids:

Carboxylic acid	Reacted with	Salt formed	Other product(s)
propanoic	potassium oxide	potassium propanoate	water
propanoic	potassium hydroxide	potassium propanoate	water
methanoic	sodium carbonate	sodium methanoate	water + carbon dioxide
methanoic	sodium	sodium methanoate	hydrogen

Table 7.5

Uses of carboxylic acids

Food preservation

As already mentioned, vinegar is a solution of ethanoic acid. It is widely used as a preservative in the food industry as the low pH of vinegar prevents bacteria and fungi growing which would spoil the food. Using vinegar in this way is an example of **pickling** (Figure 7.20). Vinegar is used to pickle eggs and various vegetables such as cucumbers.

Figure 7.20 Vinegar is used to pickle foods

Cleaning

As carboxylic acids are acidic, they can react with other substances which make them ideal for cleaning. For example, in areas of the UK where there is a high concentration of metal carbonates in the drinking water ('hard water'), metal carbonate deposits can build up in kettles, showerheads and taps. It has been found that vinegar is an ideal chemical for cleaning in such situations. Vinegar is highly effective at removing metal carbonate deposits because it contains ethanoic acid. Ethanoic acid can react with metal carbonates to produce soluble salts, water and carbon dioxide, which are washed away from the appliance being cleaned. The equation below shows the reaction between ethanoic acid and calcium carbonate:

$$2CH_3COOH(aq) + CaCO_3(s) \longrightarrow$$
$$(CH_3COO)_2Ca(aq) + H_2O(l) + CO_2(g)$$

Carboxylic acids in food

The food industry uses other carboxylic acids such as benzoic acid, C_6H_5COOH. This occurs naturally in cherry bark, raspberries and tea but it is also manufactured for use as the food additive, E210. It acts as a preservative and is often used in fruit products and soft drinks. Sodium benzoate, E211, has similar properties and is used in bottled sauces and fruit juices.

Citric acid is another carboxylic acid commonly encountered when eating food. It is found in citrus fruits, such as lemons, and is responsible for the bitter taste you get when eating these fruits. An examination of soft drink labels usually reveals the presence of citric acid as its bitter taste is an essential component of many flavoured drinks. It is also widely used in cooking. For example, fruit salads are covered in lemon juice because the citric acid stops the enzymes working in bananas and apples which would make the fruits turn brown on exposure to air.

Making medicines

Carboxylic acids can be reacted with other compounds to form medicines. For example, the painkiller aspirin can be made by reacting a carboxylic acid (salicylic acid) as shown below.

Note that both salicylic acid and aspirin contain the carboxyl group (–COOH).

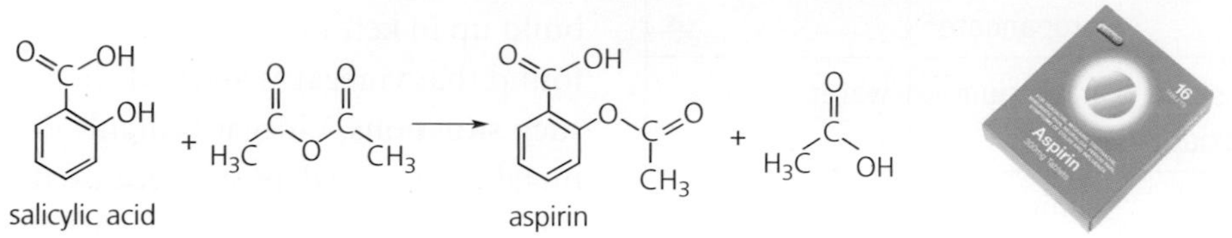

Figure 7.21 Aspirin contains the carboxyl group and can be made from the carboxylic acid, salicylic acid

Making soaps

Soaps are the sodium salts of long-chain carboxylic acids. An example is shown in Figure 7.22.

You will learn how soaps are made and how they work in the Higher Chemistry course.

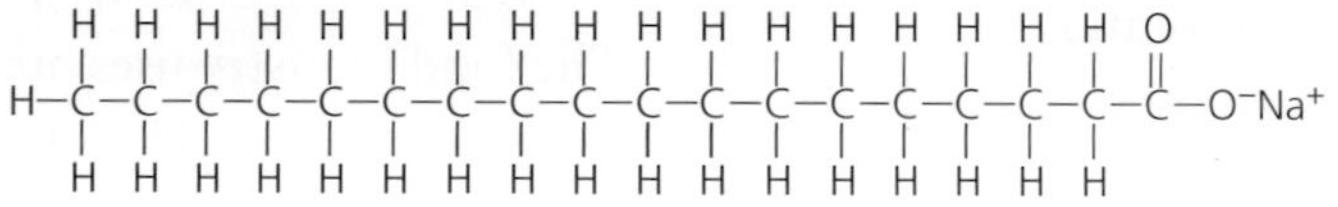

Figure 7.22 The structure of a soap

Checklist for Revision

- I can recognise the hydroxyl group, –OH, in a structural formula and know that this indicates that the compound is an alcohol.
- I can name and draw straight-chain alcohols with up to eight carbon atoms.
- I know that alcohols are very good solvents.
- I know that alcohols can be represented by the general formula $C_nH_{2n+1}OH$.
- I know that methanol, ethanol and propanol are miscible with water, and that thereafter the solubility decreases as molecular size increases.
- I know that as alcohols increase in size, their melting and boiling points increase due to the increasing strength of the intermolecular forces.
- I know that alcohols can be used as fuels as they are highly flammable and burn with very clean flames.
- I can recognise the carboxyl group, –COOH, in a structural formula and know that this indicates that the compound is a carboxylic acid.

- I can name and draw straight-chain carboxylic acids with up to eight carbon atoms.
- I know that carboxylic acids can be represented by the general formula $C_nH_{2n+1}COOH$.
- I know that methanoic, ethanoic, propanoic and butanoic acid are miscible with water, and that thereafter the solubility decreases as molecular size increases.
- I know that as carboxylic acids increase in size, their melting and boiling points increase due to the increasing strength of the intermolecular forces.
- I know that solutions of carboxylic acids have a pH less than 7 and, like other acids, can react with metal oxides, metal hydroxides, metal carbonates and metals forming salts.
- I can name the salts formed from straight-chain carboxylic acids.
- I can state some uses of carboxylic acids.

End-of-chapter questions

1 Which of the following shows the name of an alcohol with three carbon atoms?

   A Methanoic acid   B Methanol
   C Propanol   D Propanoic acid

2 Use the grid below to answer the questions that follow:

A Ethane	B Ethene
C Ethanol	D Ethanoic acid

   a) Which compound would have the lowest pH?
   b) Which compound is formed when compound B reacts with water?
   c) Which compound is found in vinegar?
   d) Which compound is formed when compound B reacts with hydrogen?
   e) Which compound can react with sodium hydroxide to form sodium ethanoate?

3 The molecule shown in Figure 7.23 is an example of
   A A saturated alcohol
   B An unsaturated carboxylic acid
   C An unsaturated alcohol
   D A saturated carboxylic acid

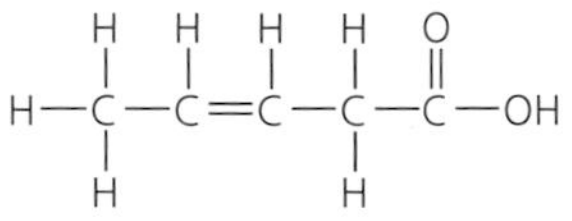

Figure 7.23

4 Draw the full structural formula for pentan-3-ol.
5 Draw the full structural formula for propanoic acid.
6 Describe two uses for carboxylic acids.
7 Ethanol can be used as a fuel for cars. Write a chemical equation for the complete combustion of ethanol.
8 Name the carboxylic acid that can react with sodium hydroxide to form the salt sodium propanoate.
9* Hydroxy acids are compounds that contain both a hydroxyl group and a carboxylic acid group within the same molecule. These compounds are able to form cyclic compounds called lactones.

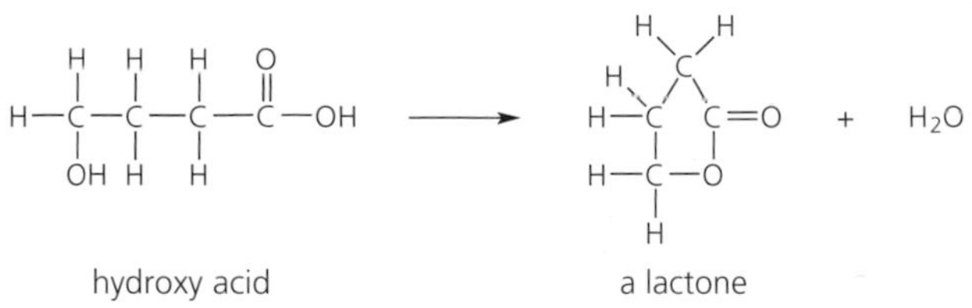

hydroxy acid   a lactone

Figure 7.24

a) Draw the structural formula for the lactone formed when this hydroxy acid reacts.

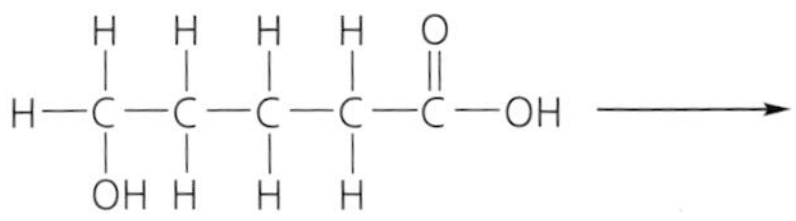

Figure 7.25

b) Draw the structural formula for the hydroxy acid from which this lactone was formed.

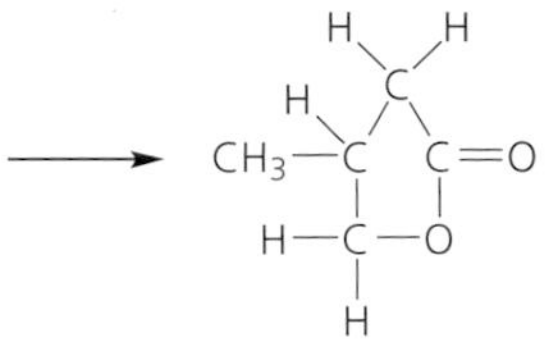

Figure 7.26

10* The structural formulae for two of the compounds in lavender oil are shown in Figure 7.27.

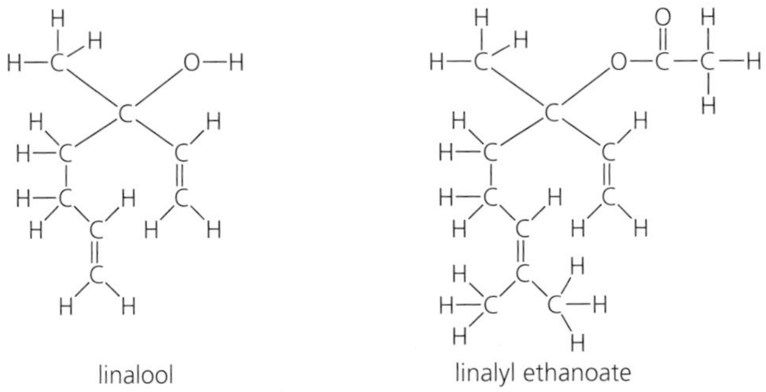

linalool   linalyl ethanoate

Figure 7.27

a) Describe how bromine solution could be used to prove that lavender oil contained unsaturated compounds.
b) Linalool can be classed as an alcohol. Draw the part of the molecule that allows you to conclude that linalool is an alcohol.
c) Draw the full structural formula for ethanoic acid.

11 Carboxylic acids can react to form salts. Complete
the following table by naming the salts formed (a–d)
and carboxylic acids reacting (e–h).

Name of carboxylic acid	Reacting with	Name of salt formed
methanoic acid	potassium oxide	a)
propanoic acid	lithium carbonate	b)
ethanoic acid	sodium hydroxide	c)
butanoic acid	sodium carbonate	d)
e)	calcium oxide	calcium hexanoate
f)	potassium hydroxide	potassium butanoate
g)	magnesium	magnesium methanoate
h)	sodium	sodium ethanoate

Table 7.6

8 Energy from fuels

Alkanes and alcohols are excellent fuels (Figures 8.1 and 8.2). When fuels burn, they release lots of energy that we can use to heat our homes or power our cars. As chemists, we want to be able to compare fuels so that we know how much energy is released from burning a certain quantity of fuel. We will examine this idea in this chapter by using our knowledge of calculations from equations and we will also find out about a simple method for calculating the actual energy released from a fuel.

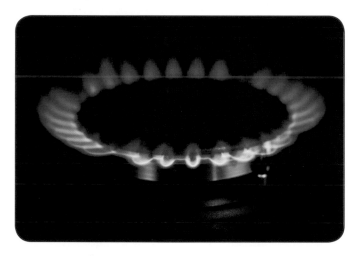

Figure 8.1 Cooking using a gas hob involves burning methane gas

Figure 8.2 Ethanol can be blended with petrol or diesel and used as a fuel for cars

Exothermic reactions

Chemical reactions that release energy are known as **exothermic reactions**. Lighting a match (Figure 8.3) is a good example of an exothermic reaction since it is obvious that energy is released in the form of heat.

Figure 8.3 Burning is an exothermic reaction

Figure 8.4 can be used to illustrate an exothermic reaction. It shows that the products have less energy than the reactants. Where has the energy gone? It has been released to the surroundings, mainly in the form of heat energy.

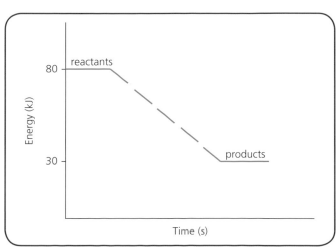

Figure 8.4 Exothermic reactions release energy to the surroundings

If we look at the values for energy on the *y*-axis, we can calculate how much energy has been released to the surroundings. The reactants had 80 kJ of energy at the start of the reaction and the products had 30 kJ at the end of the reaction. From these figures we can say that 50 kJ has been released to the surroundings. If this energy was released as heat, the temperature would rise. The word 'surroundings' can sometimes be a bit complicated so it is worth looking at an example of an exothermic reaction to understand exactly what is meant by the surroundings.

Think about adding some magnesium powder to hydrochloric acid (see Figure 8.5).

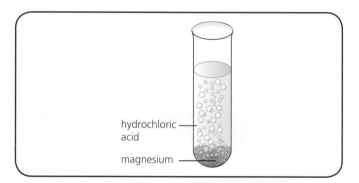

Figure 8.5 The temperature rises when magnesium is added to acid

If you have carried out this experiment in the laboratory, you will hopefully remember that the test tube becomes very warm. Here, heat energy is released to the surrounding acid, test tube and air – in other words, the surroundings are whatever is surrounding the reaction. If a thermometer was placed in the test tube, a rise in temperature would be measured.

Exothermic reactions are the most common reactions we encounter in our everyday lives as we tend to burn fuels for cooking and transport. Simple heat packs that are applied to the skin to soothe aching muscles and joints are another example of an exothermic reaction (Figure 8.6). The heat packs usually contain iron filings, salt and carbon. When exposed to air, the chemicals in the heat pack react and give out heat energy.

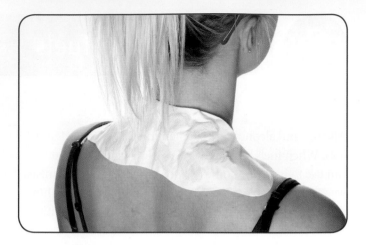

Figure 8.6 Heat packs release energy – an example of an exothermic reaction

Endothermic reactions

Some chemical reactions take in energy from the surroundings. These are known as **endothermic reactions**. For example, when barium hydroxide is mixed with ammonium chloride, the temperature falls as a reaction takes place (see Figure 8.7).

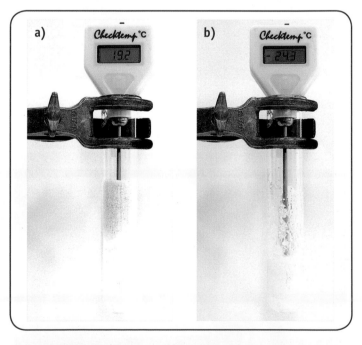

Figure 8.7 The temperature falls in an endothermic reaction: (a) 19.2°C; (b) −24.3°C

Figure 8.8 illustrates an endothermic reaction. You can see that the products have more energy than the reactants. Where has the energy come from? The energy has been taken away from the surroundings, so the surroundings become cooler. If we look at the scale on the y-axis, we can calculate that 50 kJ of energy has been removed from the surroundings. Endothermic reactions always take in energy from the surroundings.

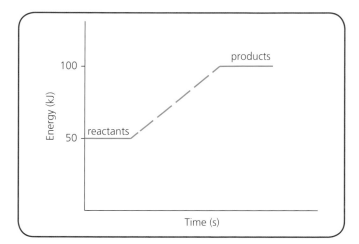

Figure 8.8 Endothermic reactions take in energy from the surroundings

Endothermic reactions are not as common as exothermic reactions, but we do find them in everyday life. For example, sherbet (Figure 8.9) contains sugar, citric acid and bicarbonate of soda. When the citric acid and bicarbonate of soda dissolve in your saliva and react with each other, there is a small drop in temperature.

Figure 8.9 Eating sherbet causes the temperature of your mouth to decrease – this is an endothermic reaction

Likewise, some cold packs (Figure 8.10) contain chemicals that react endothermically. These can be used to treat muscle sprains.

Figure 8.10 Some cold packs contain chemicals that react together to produce a drop in temperature

Writing combustion equations

Combustion (burning) is an obvious example of an exothermic reaction.

When chemicals burn they react with oxygen. For example, hydrogen gas burns readily to produce water:

$$H_2 + \tfrac{1}{2}O_2 \longrightarrow H_2O$$

When hydrocarbons burn in a good supply of oxygen, carbon dioxide and water are formed, as shown in the following equation:

ethane + oxygen $\longrightarrow$ carbon dioxide + water

$$C_2H_6 + O_2 \longrightarrow CO_2 + H_2O$$

These equations can be balanced quite easily if you follow the 'CHO' rule: balance the carbon, then the hydrogen and, finally, the oxygen atoms. For burning ethane, the balanced chemical equation is:

$$C_2H_6 + 3\tfrac{1}{2}O_2 \longrightarrow 2CO_2 + 3H_2O$$

Alcohols also burn to produce carbon dioxide and water and the CHO rule can be applied to balance alcohol combustion equations too:

methanol + oxygen $\longrightarrow$ carbon dioxide + water

$$CH_3OH + 1\tfrac{1}{2}O_2 \longrightarrow CO_2 + 2H_2O$$

Try the examples below to practise writing and balancing equations for the burning of hydrocarbons and alcohols. Review Chapter 4 if you need extra help with balancing chemical equations.

Questions

1 Write balanced chemical equations for the complete combustion of:
   a) methane
   b) butane
   c) hexane.
2 Write balanced chemical equations for the complete combustion of:
   a) ethanol
   b) propanol
   c) hexanol.

Calculations from equations

In Chapter 4, you learned about calculating quantities of reactants and products from balanced chemical equations. This skill can be applied to combustion equations where we are interested in calculating the quantities of reactants required or products formed.

Questions

3 Calculate the mass of water formed when 8 g of hydrogen is burned completely.
4 Calculate the mass of carbon dioxide formed when 60 g of ethane is burned completely.
5 Calculate the mass of water formed when 2.9 g of butane is burned completely.
6 Calculate the mass of carbon dioxide formed when 23 g of ethanol is burned completely.
7 Calculate the mass of ethanol required to produce 90 g of water when the ethanol is burned completely.
8 Calculate the mass of oxygen required to react completely with 17.6 g of propane.
9 Calculate the mass of butane required to react completely with 51.2 g of oxygen.
10 Calculate the mass of propanol required to produce 22 g of carbon dioxide on complete combustion.

Worked example

Calculate the mass of water produced when 32 g of methane is burned according to the following equation:

Step 1: Write a balanced equation, unless already given	$CH_4 + 2O_2 \longrightarrow CO_2 + 2H_2O$
Step 2: Identify the two chemicals referred to in the question and write the mole ratio	1 mole $\longrightarrow$ 2 moles
Step 3: Calculate the number of moles of the substance you have been given a mass for (methane)	$\text{number of moles} = \dfrac{\text{mass in grams}}{\text{mass of 1 mole}}$ $= \dfrac{32\,g}{16\,g}$ $= 2 \text{ moles}$
Step 4: Use the mole ratio to calculate the number of moles of the substance you are trying to find (water)	1 mole $\longrightarrow$ 2 moles 2 moles $\longrightarrow$ 4 moles
Step 5: Calculate the mass	$\text{mass in grams} = \text{number of moles} \times \text{mass of 1 mole}$ $= 4 \times 18\,g$ $= 72\,g$

Calculating the energy released from fuels

In secondary school, a simple way of comparing the energy released from burning different foods (the fuel) is to burn the food under a test tube of water and to note the starting temperature and highest temperature reached. This is shown in Figure 8.11.

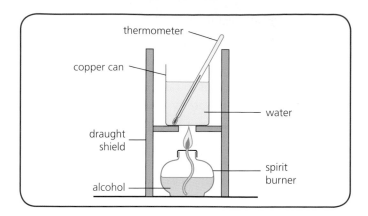

Figure 8.12 An experiment to calculate the energy released when a fuel burns

In this experiment, the fuel is burned in a spirit burner and the heat released is used to heat the water in the beaker. The mass of fuel used to heat the water can be calculated by measuring the mass of the spirit burner at the start of the experiment and at the end of the experiment. Table 8.1 shows some sample data from an experiment to calculate the energy released when ethanol burns.

Mass of burner before burning	100.0 g
Mass of burner after burning	99.5 g
Initial temperature of water	20 °C
Highest temperature of water	50 °C
Volume of water	100 cm³

Table 8.1 Experimental data

Experiment 1: Burning ethanol

In order to convert the temperature rise into an energy value, we need to use an equation that relates the temperature rise to energy. The equation we will use is:

$$E_h = cm\Delta T$$

E_h = the energy released
c = the specific heat capacity of water
m = the mass of water heated in kilograms (we assume that 1 litre of water = 1 kg)
ΔT = the temperature change in °C

The specific heat capacity of water converts the temperature rise to an energy value. It is given the

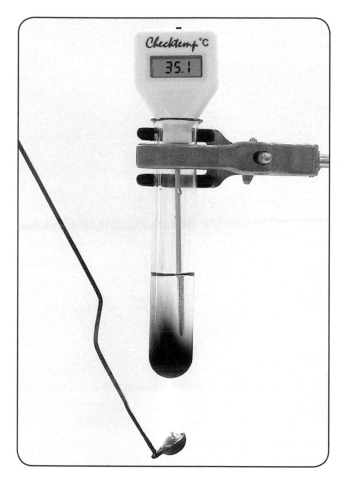

Figure 8.11 An experiment to work out the energy released when a food is burned

This experiment can provide a good comparison between different foods as long as some steps are taken to ensure that each test is fair. The volume of water heated must be the same in each experiment, the mass of food burned should be known and the distance between the burning food and test tube should be kept the same, which is quite tricky! When comparing liquid fuels, it is more convenient to use the experimental set-up shown in Figure 8.12.

symbol c and has a value of $4.18\,kJ\,kg^{-1}\,°C^{-1}$. This value tells us that to raise the temperature of 1 kg of water by 1 °C requires 4.18 kJ.

It should make sense that to raise the temperature of an even larger mass of water would require much more energy (think about using a Bunsen to boil a small volume of water versus a large volume of water). For example, to raise the temperature of 10 kg of water by 1 °C would require ten times as much energy, in other words 41.8 kJ.

Similarly, to raise the temperature of 1 kg of water by a greater amount would require more energy. For example, to raise the temperature of 1 kg of water by 2 °C would require $2 \times 4.18 = 8.36\,kJ$.

Let us apply this equation to the data in Table 8.1.

$$E_h = cm\Delta T$$

$c = 4.18$
$m = 0.1\,kg\,(100\,cm^3\,water = 100\,g = 0.1\,kg)$
$\Delta T = 50\,°C - 20\,°C = 30\,°C$
$E_h = 4.18 \times 0.1 \times 30 = 12.54\,kJ$

How much ethanol was burned? We can work this out by comparing the mass of the burner before and after burning: $100.0\,g - 99.5\,g = 0.5\,g$.

Overall we can conclude that 0.5 g of ethanol burns to release 12.54 kJ of energy.

If you were asked to calculate the energy released when, for example, 40 g of ethanol was burned, you could do this by using simple proportion:

Step 1: Write the relationship between mass and energy that you know	$0.5\,g \longrightarrow 12.54\,kJ$
Step 2: Calculate the energy for 1 g (12.54/0.5)	$1\,g \longrightarrow 25.08\,kJ$
Step 3: Calculate the energy for 40 g	$40\,g \longrightarrow$ $(40 \times 25.08\,kJ)$ $= 1003.2\,kJ$

Experiment 2: Burning methanol

An experiment was carried out to calculate the energy released when methanol burns. Use the data from Table 8.2 to show that methanol releases less energy than ethanol.

Mass of burner before burning	120.0 g
Mass of burner after burning	119.2 g
Initial temperature of water	29 °C
Highest temperature of water	54 °C
Volume of water	100 cm³

Table 8.2

$$E_h = cm\Delta T$$

$$E_h = 4.18 \times 0.1 \times 25 = 10.45\,kJ$$

Mass of methanol burned $= 120.0 - 119.2 = 0.8\,g$

In other words, 0.8 g of methanol released 10.45 kJ when burned.

$$0.8\,g \longrightarrow 10.45\,kJ$$

$$\frac{0.8\,g}{0.8} \longrightarrow \frac{10.45\,kJ}{0.8}$$

$$1.0\,g \longrightarrow 13.06\,kJ$$

By comparing the energy released when 1 g of methanol burns to the energy released when 1 g of ethanol burns (from Experiment 1), you can see that much less energy is released by burning methanol.

Improving the experiment

When we compare the energy values calculated by our simple laboratory experiment to data tables, we find that our results suggest less energy is produced. This is because a lot of the heat produced by our burning fuels is released to the surroundings and does not heat the water. Also, if you have carried out this experiment, you will have noticed that the flame from the burning fuels is often yellow and that the beaker containing the water quickly becomes coated in a layer of soot. This tells us that incomplete combustion is occurring. To improve this experiment, an instrument called a bomb **calorimeter** can be used, shown in Figure 8.13.

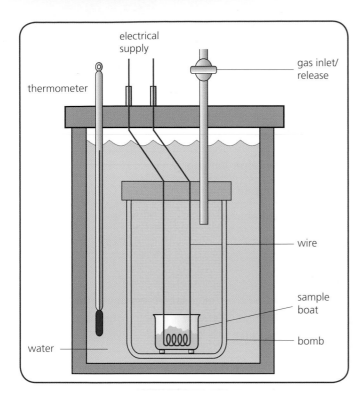

Figure 8.13 The bomb calorimeter

The calorimeter can be used to obtain much more accurate energy data as it is designed to overcome the two experimental problems that were discussed for the spirit burner:

● Heat loss to the surroundings: as the water completely surrounds the container where the combustion occurs, most of the heat generated will be transferred to the water.
● Incomplete combustion: the calorimeter is usually attached to an oxygen cylinder so that the sample is ignited in an oxygen-rich atmosphere. This ensures that the sample is completely burned.

Heating substances other than water

Experiments performed to calculate the quantity of heat released by a fuel will involve measuring the temperature rise in a known quantity of water. In other experiments, the heat released by a chemical reaction can result in substances other than water heating up. For example, adding an acid to an alkali is an example of an exothermic reaction. To calculate the heat released, the equation:

$$E_h = cm\Delta T$$

can still be used.

In this case:

● c is the specific heat capacity of water
● m is the mass of solution being heated (taken as the mass of the acid + alkali)
● ΔT is the difference between the highest temperature reached by the mixture and the average starting temperature of the acid and alkali.

It may seem unusual that the specific heat capacity of water is used in this case as the heat released is being absorbed by a salt solution (since a salt is formed when an acid is reacted with an alkali). However, since most of the solution is water, we can use $c = 4.18$ kJkg^{-1} °C^{-1} as an approximation for the specific heat capacity of the salt solution.

Scientists also perform experiments in which the heat released is absorbed by substances other than water or an aqueous solution. For example, they may wish to calculate the energy required to heat a length of metal to a desired temperature. The equation $E_h = cm\Delta T$ can still be used, however the value for c will not be 4.18. Instead, c will be the specific heat capacity of the substance being heated, in this case the metal.

Checklist for Revision

● I can state the definitions of the words 'exothermic' and 'endothermic' and I know that burning is an example of an exothermic reaction.
● I can write balanced chemical equations for the burning of hydrocarbons and alcohols.
● I can calculate quantities of reactants and products using chemical equations.
● I can calculate the energy released from a fuel using $E_h = cm\Delta T$.

End-of-chapter questions

1 For each of the following examples, state whether the reaction is exothermic or endothermic and calculate the energy released/taken in.
   a) A chemical reaction took place where the reactants had 80 kJ of energy. At the end of the experiment, the products were found to have 20 kJ of energy.
   b) A chemical reaction took place where the reactants had 120 kJ of energy. At the end of the reaction, the products were found to have 190 kJ of energy.

2 Calculate the mass of water formed when 20 g of hydrogen is completely burned.

3 Calculate the mass of carbon dioxide formed when 640 g of methane burns completely.

4 Calculate the mass of hexane required to produce 144 g of water on complete combustion.

5 Calculate the mass of carbon dioxide formed when 10 g of methanol is burned completely.

6 Calculate the mass of ethanol required to react completely with 100 g of oxygen.

7* The energy released by burning methanol can be calculated using the apparatus shown in Figure 8.14. In an experiment, the results given in Figure 8.15 were obtained. Use these results to calculate the energy released when 32 g of methanol is burned.

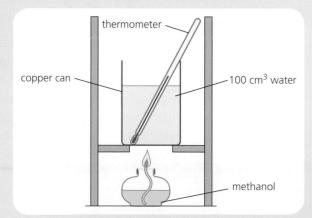

Figure 8.14

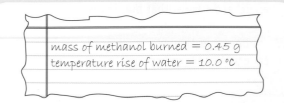

mass of methanol burned = 0.45 g
temperature rise of water = 10.0 °C

Figure 8.15

8 Use the data in Table 8.3 to calculate the energy released when 100 g of propanol is completely burned using the apparatus shown in question 7.

Mass of burner before burning	120.0 g
Mass of burner after burning	119.2 g
Initial temperature of water	20 °C
Highest temperature of water	57 °C
Volume of water	100 cm³

Table 8.3

9* A student used the simple laboratory apparatus shown in Figure 8.16 to determine the energy released when methanol is burned.

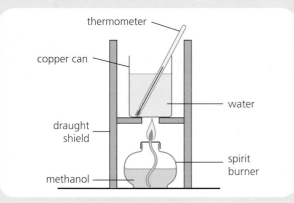

Figure 8.16

a) What measurements would have to be made to allow you to calculate the energy released by burning the methanol?
b) The student found that burning 0.370 g of methanol produces 3.86 kJ of energy. Calculate the energy released when 20 g of methanol is burned.
c) A more accurate value for the energy released by burning a fuel can be obtained using a bomb calorimeter (Figure 8.17). From your reading in this chapter, suggest why this experiment gives more accurate results than the simple spirit burner method.

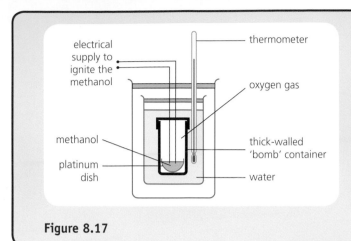

electrical supply to ignite the methanol

thermometer

oxygen gas

methanol

thick-walled 'bomb' container

platinum dish

water

Figure 8.17

10* Scientists have been experimenting to find ways of reducing carbon dioxide in the atmosphere. One of these ways involves placing concrete balls on the sea bed. They hope that green plants called algae will grow on the balls and this will help to reduce the carbon dioxide level.

The balanced equation for the removal of carbon dioxide by algae is:

$$6CO_2 + 6H_2O \longrightarrow C_6H_{12}O_6 + 6O_2$$

If 200 kg of carbon dioxide were removed by the algae, calculate the mass of oxygen that would be produced.

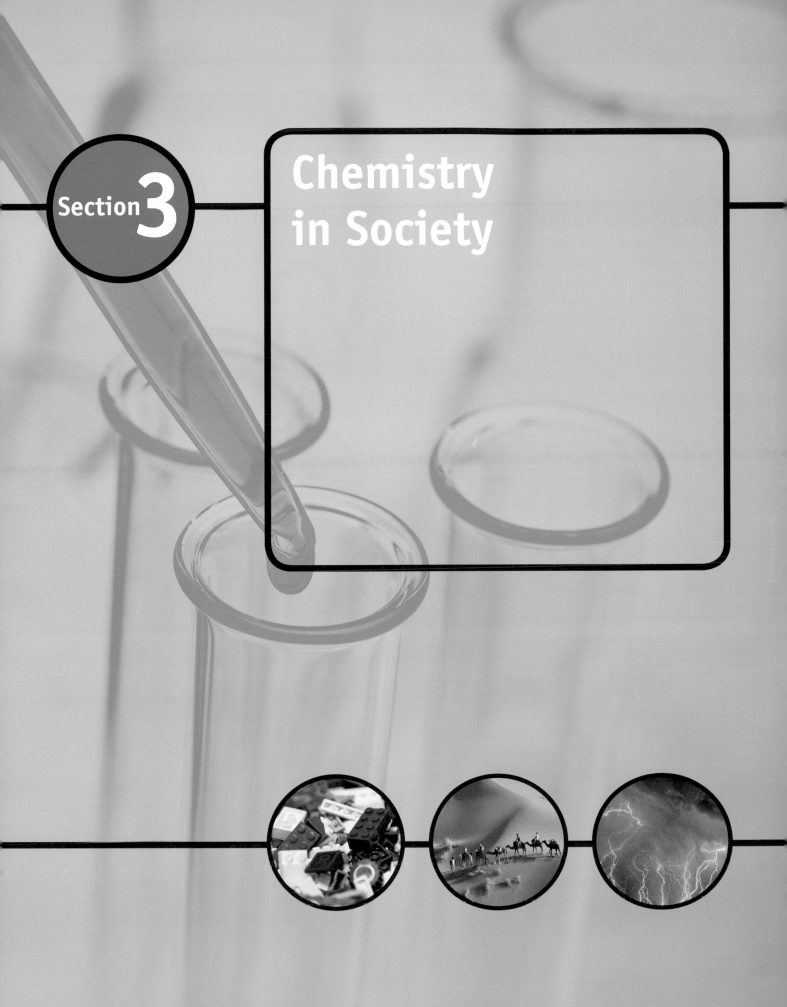

Section **3**

Chemistry in Society

9   Metals

Crystals and bonding

In Chapter 3, you learned that ionic bonds are the electrostatic forces of attraction between positive and negative ions, and that these electrostatic attractions cause the oppositely charged ions to assemble themselves into a regular, three-dimensional structure called an ionic lattice (Figure 9.1). The arrangement of ions in the lattice is a balance between the repulsion experienced by ions of the same charge and the attraction between ions of opposite charge.

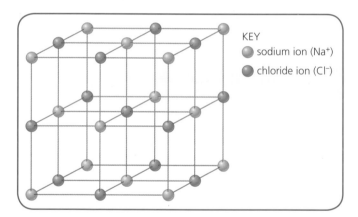

KEY
- sodium ion (Na⁺)
- chloride ion (Cl⁻)

Figure 9.1 An ionic lattice

Common salt contains sodium ions (Na^+) and chloride ions (Cl^-). Both of these ions have a fixed size (sodium ions are about half the size of chloride ions), so the ions assemble in a particular pattern; a pattern that repeats itself over and over again. This pattern is so regular that as more and more ions join the lattice, **macroscopic** particles of a fixed shape (decided by the pattern) start to form and these become big enough for us to see. We call them crystals. Many substances that consist of a regular, three-dimensional array of particles can form crystals. For example, carbon atoms can arrange themselves into an **atomic lattice** and form crystals that we call diamonds. Odd though it may seem, metals also have a crystal structure.

Metallic bonding

Figure 9.2 We use metals in many ways

Metal atoms have a small number of outer electrons, so in order to achieve the electron arrangement of their nearest noble gas, metal atoms tend to lose electrons. When a metal atom loses its outer electron(s), a positive ion is formed. For example:

$$Mg \longrightarrow Mg^{2+} + 2e^-$$

(neutral   (positive   (two electrons lost
atom)   ion)   by the Mg atom)

If we could see into the structure of a metallic element, we would see that all the metal atoms had lost their outer electrons and had formed a regular, three-dimensional array (lattice) of positively charged ions. These positive ions should all repel each other and blow the structure apart, so why does this not happen? The reason is that the electrons lost from the outer energy level of each atom remain in the lattice and move freely between the positive ions. This 'sea' of negative charge (electrons) floating through the metallic lattice does two things:

- It attracts the positive ions (holding them together).
- It balances the repulsion between the positive ions (stopping them being forced apart).

The free electrons in metals are usually said to be delocalised because they are not held to any one nucleus; they can move anywhere through the structure and the electrostatic attraction of the positive ions for the delocalised electrons is called **metallic bonding** (Figure 9.3).

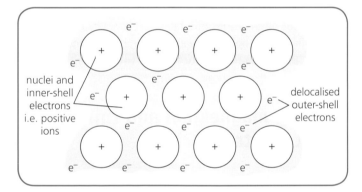

Figure 9.3 Metallic bonding

Metallic bonding is usually strong, so most metal elements have got high melting points and all but one of the metal elements exist in the solid state at room temperature (25 °C).

Properties of metals

Ionic lattices and metallic lattices both contain charged particles, but in an ionic lattice, the oppositely charged ions in the compound have different sizes and different charges, so they cannot change places without destroying the pattern of the lattice. A metal, however, is an element, so in a pure metal, the positive ions are all identical; they are all the same size and they can change places in the lattice and slide over each other without breaking the structure apart. This gives metals some very useful properties. For instance, if you pick up a small copper(II) sulfate crystal (taking care to wear gloves)

and try to bend it, it will break, but if you do the same to a small piece of metal foil, it will not break; it will bend. We say metals are **malleable**, which means they can be shaped by beating them with a hammer. This property has allowed craftsmen to make tools (and swords!) from metal for thousands of years.

Metallic bonding also makes metal elements very good conductors of electricity. Electricity is a flow of charged particles and, if a voltage is applied across a piece of metal, the free (delocalised) electrons can move easily through the lattice. As a result, an electric current can flow through the metal without any chemical change taking place. Metals and graphite both conduct electricity in the solid state because their structures contain delocalised electrons. Due to their type of bonding, metals can conduct electricity when they are solid and when they are **molten** (in the liquid state) and they always remain chemically unchanged by the electron flow.

It is important to remember that although the delocalised electrons in a metal are no longer held by one particular nucleus, they have not been transferred to another type of atom; the metal has not reacted with anything and the outer electrons from each nucleus are all still present in the lattice. This means that although the lattice is an array of positive ions, the metal element consists of atoms. The symbol for the element magnesium is Mg. The symbol Mg^{2+} is only used when magnesium atoms have reacted with another type of atom and the element, magnesium, has become part of a compound.

Activities

Farriers make iron horseshoes to attach to horses' hooves. They use a hammer to beat the hot metal into shape. Use your knowledge of chemistry to write a short paragraph describing what happens to the metal lattice when a horseshoe is being forged (being made) and explain why the piece of iron does not break when the hammer hits it.

Reactions of metals

Metal atoms all have relatively small numbers of outer electrons, so they have one chemical property in common: when they take part in a chemical reaction, they tend to lose electrons. We often say that metal atoms tend to give electrons away or 'donate' electrons. The atoms of some metals lose (donate) electrons more easily than others. We say that metals whose atoms lose electrons easily are reactive metals and metals whose atoms do not lose electrons easily are unreactive metals. Due to this tendency to give electrons away, all metals have similar reactions with air (oxygen), water and with dilute acids. It is possible to put together a reactivity series of metals based on their reactions with oxygen, water and dilute acids. This series is shown in Table 9.1.

Name	Symbol
potassium	K
sodium	Na
lithium	Li
calcium	Ca
magnesium	Mg
aluminium	Al
zinc	Zn
iron	Fe
tin	Sn
lead	Pb
copper	Cu
mercury	Hg
silver	Ag
gold	Au

Table 9.1 Some metals are more reactive than others. This table gives a reactivity series for some common metals, with the most reactive metals at the top and the least reactive at the bottom

Burning magnesium and thinking about electrons

The balanced equation for the combustion of magnesium is:

$$2Mg(s) + O_2(g) \longrightarrow 2MgO(s)$$

When metals take part in a chemical reaction, they lose electrons. In this reaction, the electrons lost (as the magnesium atoms become magnesium ions, Mg^{2+}) must be gained by the atoms in the oxygen molecule, which means the oxygen atoms become oxide ions, O^{2-}.

Equation 1: $2Mg(s) + O_2(g) \longrightarrow 2Mg^{2+}O^{2-}(s)$

We can split this equation into two halves; one half shows what happens when a magnesium atom (electron arrangement 2,8,2) loses its two outer electrons:

Equation 2: $Mg \longrightarrow Mg^{2+} + 2e^-$

The other half shows what happens when an atom from the oxygen molecule (the electron arrangement of an oxygen atom is 2, 6) gains the electrons lost from the magnesium atom:

Equation 3: $\frac{1}{2}O_2 + 2e^- \longrightarrow O^{2-}$

Equations 2 and 3 each represent half of the reaction shown in equation 1 and since they contain both ions and electrons, they are called ion–electron half equations (often simply called ion–electron equations). When magnesium burns, it combines with oxygen and we say that it has been oxidised. Equation 2 shows that when magnesium burns, its atoms lose electrons, so we can also say that when a metal atom loses electrons, it is being oxidised. In chemistry, the opposite of **oxidation** is **reduction**, so we say the oxygen atoms in equation 3 are being reduced, in other words they are gaining electrons. An easy way to remember how the electrons move in these two processes is to use the mnemonic OIL RIG:

Oxidation **I**s **L**oss (of electrons); **R**eduction **I**s **G**ain (of electrons)

Remember: in an ion–electron equation showing oxidation (often indicated by writing [O] following it), the electrons are moving away (being lost) from the reactant, so the electrons appear on the *right* of the arrow, but in an ion–electron equation showing reduction (often indicated by writing [R] following it), the electrons are moving towards (being gained by) the reactant, so the electrons appear on the *left* of the arrow.

Reaction of metals with oxygen

All the metals above silver in the reactivity series (see Table 9.1) combine with oxygen when heated. The reaction produces a metal oxide. The higher the metal is in the series, the more violent the reaction between the metal and oxygen. In general, this reaction takes the form:

$$metal + oxygen \longrightarrow metal\ oxide$$

For example,

$$copper + oxygen \longrightarrow copper(II)\ oxide$$
$$2Cu + O_2 \longrightarrow 2CuO$$

Metal oxides are ionic compounds, so have high melting points and are solids at room temperature.

Showing ions and state symbols, the equation becomes:

$$2Cu(s) + O_2(g) \longrightarrow 2Cu^{2+}O^{2-}(s)$$

The apparatus shown in Figure 9.4 can be used to study the reactivity of metals from magnesium down to copper in the reactivity series.

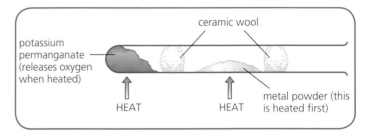

Figure 9.4 Metals glow brightly as they react with oxygen

The three metals at the top of the reactivity series – potassium, sodium and lithium – are so reactive that they are stored under oil to prevent them from reacting with the oxygen and water in the air.

Reactions of metals with water

The metals above aluminium in the reactivity series react with cold water to produce hydrogen gas and the corresponding metal hydroxide. The general equation for this is:

$$metal + water \longrightarrow metal\ hydroxide + hydrogen$$

For example:

$$sodium + water \longrightarrow sodium\ hydroxide + hydrogen$$
$$2Na + 2H_2O \longrightarrow 2NaOH + H_2$$

If we rewrite this equation to show ions and state symbols, we get:

$$2Na(s) + 2H_2O(l) \longrightarrow 2Na^+(aq) + 2OH^-(aq) + H_2(g)$$

Reactions of metals with dilute acids

All metals above copper in the reactivity series react with dilute acids such as hydrochloric acid and sulfuric acid to produce a salt and hydrogen:

$$metal + acid \longrightarrow salt + hydrogen$$

For example:

zinc + hydrochloric acid $\longrightarrow$ zinc(II) chloride + hydrogen

$$Zn \ + \quad 2HCl \quad \longrightarrow \quad ZnCl_2 \ + \ H_2$$

If we rewrite this equation to show ions and state symbols, we get:

$$Zn(s) + 2H^+(aq) + 2Cl^-(aq) \longrightarrow$$
$$Zn^{2+}(aq) + 2Cl^-(aq) + H_2(g)$$

When a metal reacts with an acid it produces bubbles of hydrogen gas. In general, the faster the bubbles are produced, the more reactive the metal. One exception is aluminium. It reacts slowly with acids for about the first 20 minutes, after which it reacts quickly. The reason for this is that the aluminium is protected by a coating of aluminium oxide, which must be removed by the acid before the metal can react. Table 9.2 summarises the reactions of metals.

Using a metal/acid reaction to make a salt

As you saw in Chapter 5, a salt is made when the hydrogen ions in an acid are replaced by metal ions or (as we will see in Chapter 11 on Fertilisers) by ammonium ions. In Chapter 5, salts were produced by the neutralisation reaction of an acid by a base. The reaction of an acid with a metal does produce a salt but it does not produce water so it is not a neutralisation reaction.

If we want to use this type of reaction to make a particular salt, we have to think carefully about the reactivity of the metal involved. Metals high up in the reactivity series, the ones like potassium, sodium, lithium and calcium that react readily with cold water, are so reactive that adding them to dilute acids produces a dangerously exothermic reaction. Magnesium only reacts very slowly with cold water so it can safely be added to dilute acids, for example

$$metal \ + \quad acid \quad \longrightarrow \quad salt \quad + \ hydrogen$$

magnesium + hydrochloric $\longrightarrow$ magnesium + hydrogen
            acid   chloride

Metal	Reaction with		
	Oxygen	Water	Dilute acid
potassium	metal + oxygen ↓ metal oxide	metal + cold water ↓ metal hydroxide + hydrogen	metal + acid ↓ salt + hydrogen
sodium			
lithium			
calcium			
magnesium			
aluminium		no reaction	
zinc			
iron			
tin			
lead			
copper			no reaction
mercury			
silver	no reaction		
gold			

Table 9.2

Procedure

A measuring cylinder is used to pour $10\,cm^3$ of dilute hydrochloric acid into a $250\,cm^3$ beaker. Magnesium powder or small pieces of magnesium ribbon are added to the acid, and the liquid is stirred until adding a little more metal no longer makes it fizz. The mixture is filtered, the filtrate is poured into an evaporating basin and heated until the liquid has all evaporated and only solid remains; this is called 'evaporating to dryness'.

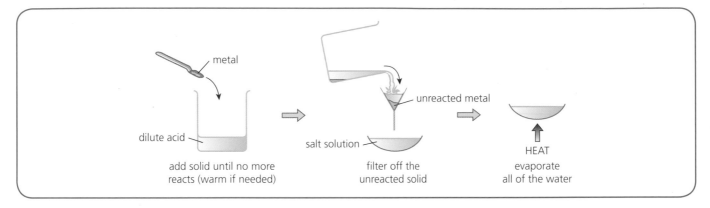

Figure 9.5 Preparing soluble salts using a metal

Supply and demand

Humans have been digging up metal ores and extracting the metals for thousands of years. A metal ore is a rock that contains a metal compound (miners call metal compounds 'minerals') and a good knowledge of chemistry is needed to get (extract) the pure metal from each metal compound. A lot of mines are now worked out, which means that the amount of mineral left in the mine is not worth the cost of digging it up, and the big mining companies around the world are constantly looking for new supplies. We know about the demand for metals such as iron, aluminium and copper because we see them in use every day, but there is a group of elements called the rare earth metals which are in high demand because they are essential for the manufacture of computers, mobile phones and GPS satellites. China currently supplies around 90% of the world production of rare earth metals but, if we do not conserve supplies, we may run out of the raw materials needed for our complex modern technologies.

Figure 9.6 Modern boats use hi-tech radar equipment

Questions

4 a) Write a balanced equation for the reaction between potassium and oxygen.
   b) Write an equation showing the ions present in the reaction. Use the equations given in the text as a guide.
5 For the reaction between calcium and water, give:
   a) a balanced equation
   b) an equation showing ions and state symbols.
6 For the reaction between magnesium and dilute hydrochloric acid, give:
   a) a balanced equation
   b) an equation showing ions and state symbols.
7 A, B and C are three metals. Metal A reacts with dilute hydrochloric acid but not with water. Metal B does not react with water or dilute acid. Metal C reacts with water and dilute acid.
   a) Place the metals in order of reactivity, with the most reactive first.
   b) Using the reactivity series in Table 9.2, give all the possible metals that A, B and C could be.
   You may need to carry out some internet research to answer Questions 8 and 9.
8 The semiconductor gallium nitride is essential for high-power radar systems (see Figure 9.6), so trade in gallium nitride is restricted.
   a) Which group of the Periodic Table is gallium in?
   b) What is the formula for gallium nitride?
   c) Name another material that is more commonly used as a semiconductor.
9 Name two rare earth metals and give the symbol for each of them.

Extraction of metals

Gold (Figure 9.7) is a very unreactive metal (it tends not to form compounds with other elements) so it can still be found as an element in the Earth's crust. Gold is almost always found as a natural alloy, mixed with other elements, usually silver. There are some deposits of other uncombined metals around the world, but most metals are obtained from their naturally occurring compounds.

Figure 9.7 Gold jewellery that dates from the Bronze Age

Industrial chemists have developed complex ways to extract some metals, but the basic chemistry used to extract the more common metals has been unchanged for hundreds, if not thousands, of years. The first step is to crush the ore into small pieces, then heat it in air to form the metal oxide. The oxygen is then removed by a process called reduction and the metal is obtained.

The reactivity series shown in Table 9.1 can be used to decide how easy it is to extract a metal from its oxide. Metals towards the bottom of the series (unreactive metals) do not form compounds easily, so it does not require much energy to break up (**decompose**) their oxides and get the pure metal. Metals towards the top of the series (reactive metals) form compounds easily, so it requires a great deal of energy to extract those metal elements from their oxides. For this purpose, we can split the reactivity series at two points producing three groups of metals.

1 Metals below copper

The oxides of these metals decompose when they are heated to give the metal and oxygen.

$$metal\ oxide \longrightarrow metal + oxygen$$

For example:

$$silver(I)\ oxide \longrightarrow silver + oxygen$$

$$2Ag_2O \longrightarrow 4Ag + O_2$$

Showing ions and state symbols, the equation becomes:

$$2(Ag^+)_2O^{2-}(s) \longrightarrow 4Ag(s) + O_2(g)$$

2 Metals above mercury and below aluminium

The oxides of these metals do not decompose when they are heated, but if the oxide is mixed with carbon and then heated, the carbon and oxygen combine to give carbon dioxide and leave the metal as an element. This removal of oxygen is called reduction. Carbon is doing the reducing (removing oxygen) so carbon is described as a **reducing agent**.

$$metal\ oxide + carbon \longrightarrow metal + carbon\ dioxide$$

For example:

$$zinc(II)\ oxide + carbon \longrightarrow zinc + carbon\ dioxide$$

$$2ZnO + C \longrightarrow 2Zn + CO_2$$

Showing ions and state symbols, the equation becomes:

$$2Zn^{2+}O^{2-}(s) + C(s) \longrightarrow 2Zn(s) + CO_2(g)$$

Another reducing agent used in the extraction of metals is carbon monoxide:

$$lead(II)\ oxide + carbon\ monoxide \longrightarrow lead + carbon\ dioxide$$

$$PbO(s) + CO(g) \longrightarrow Pb(s) + CO_2(g)$$

Carbon monoxide is a poisonous gas; it is mainly used in the extraction of iron from its ore because the gas can be both produced and used inside a large, closed vessel called a blast furnace:

$$iron(III)\ oxide + carbon\ monoxide \longrightarrow iron + carbon\ dioxide$$

$$Fe_2O_3(s) + 3CO(g) \longrightarrow 2Fe(s) + 3CO_2(g)$$

3 Metals above zinc

Compounds of these metals are very stable (hard to decompose), so a large amount of electrical energy is needed to obtain each metal. Metals above zinc in the reactivity series can only be obtained by **electrolysis** of their molten (melted) compounds. Electrolysis is a process in which ionic compounds are decomposed into simpler substances using electricity. Electrolysis depends on the movement of ions, but in solid ionic compounds the ions are held in a lattice (as you learned in Chapter 3), so they cannot move. Heating the solid gives the ions more energy and, as the solid melts, the ions become free to move and the technique of electrolysis can be applied.

During electrolysis, direct current (d.c.) electricity supply must be used so that one electrode stays negatively charged and the other stays positively charged. Most reactive metals are obtained by electrolysis of their molten chlorides; for example calcium can be obtained from calcium chloride, $Ca^{2+}(Cl^-)_2$.

In Chapter 2, you learned that atoms of metal elements form positive ions. Negative attracts positive, so positive metal ions move to the negative electrode where they pick up electrons and change from being charged ions

to being neutral atoms. We say that metal ions are 'discharged' at the negative electrode. This is a reduction reaction.

At the negative electrode:

$$Ca^{2+} + 2e^- \longrightarrow Ca \quad [R]$$

Atoms of non-metal elements form negative ions. Positive attracts negative, so negative non-metal ions move to the positive electrode where they give up electrons and change from being charged ions to being neutral atoms. We say that non-metal ions are 'discharged' at the positive electrode. This is an oxidation reaction. If the non-metal element exists as diatomic molecules, pairs of atoms then combine to form molecules.

At the positive electrode:

$$2Cl^- \longrightarrow Cl_2 + 2e^- \quad [O]$$

The overall reaction can be shown as

$$Ca^{2+}(Cl^-)_2(l) \xrightarrow{\text{electrolysis}} Ca(s) + Cl_2(g)$$

Aluminium is one of the most widely used metals in the world and it is produced by the electrolysis of its molten oxide, $(Al^{3+})_2(O^{2-})_3$.

Activities

1 Use the internet to find out more about the extraction of aluminium from its ore. Think about the following questions:
   - What is the name of the main ore of aluminium?
   - How is the metal extracted?
   - Why is the ore mixed with cryolite?
   - What material is used to make the electrodes?
   - Why does the positive electrode need to be replaced more often than the negative electrode?

2 Since the start of the twentieth century, aluminium has been relatively cheap and plentiful, so why did Napoleon Bonaparte have the state dinner plates made from aluminium as a sign of his great wealth?

Questions

10 For the effect of heat on mercury(II) oxide
   a) write a balanced equation
   b) write an equation showing ions and state symbols.
11 Copper can be obtained by heating copper(II) oxide with carbon. For this reaction:

a) write a balanced equation
b) write an equation showing ions and state symbols
c) name the reducing agent.

Reduction, oxidation and redox reactions

Whenever a reaction involves both oxidation and reduction, it is called a **redox reaction**. In a redox reaction, electrons lost by one substance during oxidation are gained by another substance during reduction. Three examples of redox reactions involving metals are their reactions with water, oxygen and dilute acids.

Metal + water

Sodium reacts readily with water according to the equation:

sodium + water $\longrightarrow$ sodium hydroxide + hydrogen

$$2Na(s) + 2H_2O(l) \longrightarrow 2Na^+OH^-(aq) + H_2(g)$$

We can split this redox equation up into ion–electron equations to show how electrons are transferred in the redox reaction. Sodium atoms are oxidised (lose electrons):

oxidation part: $Na(s) \longrightarrow Na^+(aq) + e^-$

Hydrogen ions present in water are reduced (gain the electrons lost by the sodium atoms):

reduction part: $2H^+(aq) + 2e^- \longrightarrow H_2(g)$

From the balanced equation we can see that two sodium atoms reacted. This was to produce the two electrons required to reduce hydrogen ions to hydrogen gas, so we can write equations to show the number of electrons transferred:

$$2Na(s) \longrightarrow 2Na^+(aq) + 2e^- \text{ [O]}$$

$$2H^+(aq) + 2e^- \longrightarrow H_2(g) \text{ [R]}$$

So the redox equation for this reaction would be:

$$2Na(s) + 2H^+(aq) \longrightarrow 2Na^+(aq) + H_2(g)$$

Note that this redox equation does not include the hydroxide ions present in water (reactant) or the hydroxide ions of the sodium hydroxide (product). Redox equations only show the substances that lost and gained electrons.

Metal + oxygen

Calcium reacts readily with oxygen molecules in the atmosphere according to the equation:

calcium + oxygen $\longrightarrow$ calcium oxide

overall reaction:

$$2Ca(s) + O_2(g) \longrightarrow 2Ca^{2+}O^{2-}(s)$$

We can split this redox equation up into ion–electron equations to show how electrons are transferred in the redox reaction. Calcium atoms are oxidised (lose electrons):

oxidation part: $Ca(s) \longrightarrow Ca^{2+}(s) + 2e^-$

Oxygen atoms are reduced (gain the electrons lost from the calcium atoms):

reduction part: $\frac{1}{2}O_2(g) + 2e^- \longrightarrow O^{2-}(s)$

From the balanced equation we can see that two calcium atoms are needed to react with each oxygen molecule, so we can write equations to show the number of electrons transferred in each reaction:

$$2Ca(s) \longrightarrow 2Ca^{2+}(s) + 4e^- \text{ [O]}$$

$$O_2(g) + 4e^- \longrightarrow 2O^{2-}(s) \text{ [R]}$$

Note: If the ion–electron equations for a reaction contain equal numbers of electrons, they are simply added together to give the overall redox equation with the electrons omitted.

Redox equations must *never* contain electrons. The following is an incorrect redox equation:

$$2Ca(s) + O_2(g) + 4e^- \longrightarrow 2Ca^{2+}O^{2-}(s) + 4e^-$$

The electrons lost and gained must be left out to give the following correct redox equation:

$$2Ca(s) + O_2(g) \longrightarrow 2Ca^{2+}O^{2-}(s)$$

Metal + dilute acid

The reaction between a metal and a dilute acid is a redox reaction because the metal atoms lose electrons to form metal ions and the hydrogen ions, present in the acid, gain electrons to form hydrogen gas. For example, magnesium reacts with dilute hydrochloric acid:

magnesium + hydrochloric acid $\longrightarrow$
                    magnesium chloride + hydrogen

$$Mg + 2HCl \longrightarrow MgCl_2 + H_2$$

Equation showing ions and state symbols:

overall reaction:

$$Mg(s) + 2H^+(aq) + 2Cl^-(aq) \longrightarrow$$
$$Mg^{2+}(aq) + 2Cl^-(aq) + H_2(g)$$

oxidation part: $Mg(s) \longrightarrow Mg^{2+}(aq) + 2e^-$

reduction part: $2H^+(aq) + 2e^- \longrightarrow H_2(g)$

If we add these two ion–electron equations and cancel the electrons, we get the following redox equation:

$$Mg(s) + 2H^+(aq) \longrightarrow Mg^{2+}(aq) + H_2(g)$$

There are no chloride ions in this equation because the chloride ions, $Cl^-(aq)$, are neither oxidised nor reduced during the reaction. The chloride ions are present in the mixture, but they play no part in the reaction. They are said to be spectator ions (see Chapter 5). Redox equations only show the substances that are involved in the oxidation part and the reduction part of the reaction; spectator ions are never included in redox equations.

Activities

Your teacher/lecturer may demonstrate the experiment shown in Figure 9.8. Use the electrochemical series from the Data Booklet to help you write ion–electron equations for the formation of lead metal at the negative electrode and the formation of bromine at the positive electrode. Add these two equations together and write the overall equation for the decomposition of lead(II) bromide. Explain how these equations tell you that electrolysis is a redox process. Are there any spectator ions in this reaction?

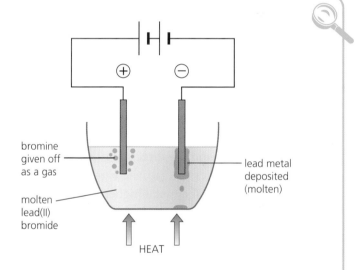

bromine given off as a gas

lead metal deposited (molten)

molten lead(II) bromide

HEAT

Figure 9.8 Passing a direct current through molten lead(II) bromide

Questions

12 State whether the following are oxidation or reduction reactions:
   a) $Na \longrightarrow Na^+ + e^-$
   b) $I_2 + 2e^- \longrightarrow 2I^-$
   c) $S + 2e^- \longrightarrow S^{2-}$.

13 Identify the redox reaction shown in the grid.

A	B
$2F^- \longrightarrow F_2 + 2e^-$	$Fe + Pb^{2+} \longrightarrow Fe^{2+} + Pb$
C	D
$e^- + K^+ \longrightarrow K$	$2O^{2-} \longrightarrow O_2 + 4e^-$

14 In the first stage of rusting, iron atoms lose electrons to form iron(II) ions:

Equation 1: $Fe \longrightarrow Fe^{2+} + 2e^-$

The electrons lost by the iron are gained by water and oxygen molecules:

Equation 2: $2H_2O(l) + O_2(g) + 4e^- \longrightarrow 4OH^-(aq)$

   a) Copy the ion–electron equations 1 and 2 and indicate which one represents reduction [R] and which represents oxidation [O].
   b) Adjust the number of electrons lost and gained in equations 1 and 2, then add the two ion–electron equations to obtain a balanced redox equation.

Electrochemical cells

In the section on 'Extraction of metals', we looked at the electrolysis of some molten compounds. Electrolysis uses electrical energy (usually from the mains) to make a chemical reaction take place, i.e. during electrolysis, electrical energy is converted into chemical energy. Scientists have found a way to reverse this reaction. An electrochemical cell is a piece of equipment in which a chemical reaction produces an electric current. An arrangement of two or more cells connected together is called a **battery**, so the development of electrochemical cells has changed our lives by giving mobile communication to the world. However, in this section it is important to remember that there is a big difference between electrochemical reactions and electrolysis reactions.

When two different metals are connected by an **electrolyte**, an electric current flows from one metal to the other through connecting wires (Figure 9.10). Comparing pairs of metals allows us to construct a list called the **electrochemical series**, which can be used to predict the size of the voltage between any two connected metals and the direction of current in chemical cells. The bigger the gap, the bigger the voltage and electrons flow from the metal higher in the series to the metal lower in the series. The electrochemical series can be found in the data Data Booklet.

Figure 9.11 shows that electrons are being transferred through the wires (often called the 'external circuit') from zinc (the metal higher in the electrochemical series) to copper (the metal lower in the electrochemical series). This means that in the zinc half-cell, zinc atoms are being oxidised to form ions, and in the copper half-cell, copper ions are being reduced to form atoms. Ion–electron equations for these reactions are:

oxidation: $Zn(s) \longrightarrow Zn^{2+}(aq) + 2e^-$

reduction: $Cu^{2+}(aq) + 2e^- \longrightarrow Cu(s)$

These two equations contain equal numbers of electrons so they can be added (remembering to omit the electrons) to give the redox equation:

$$Zn(s) + Cu^{2+}(aq) \longrightarrow Zn^{2+}(aq) + Cu(s)$$

As Figure 9.11 shows, electrons flow through the external circuit (the wires) from the metal that is losing electrons (the oxidation reaction) to the metal in the other half-cell where the reduction reaction takes place. This makes it quite easy to work out the direction of flow of electrons in a cell that does not contain metal elements; they always flow from the half-cell where the oxidation reaction takes place to the half-cell where the reduction reaction takes place. This also means that if we know the direction of flow of electrons in a cell, we immediately know which half-cell contains the oxidation reaction and which contains the reduction reaction.

N.B. the ion bridge (often called a salt bridge) is used to link the half-cells; ions move across this bridge to complete the electrical circuit.

Figure 9.9 Handsets such as this are battery operated

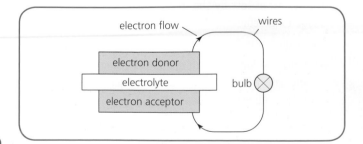

Figure 9.10 How an electrochemical cell works

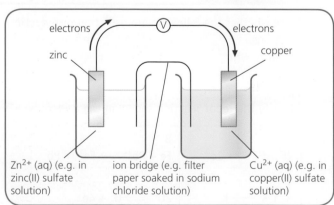

Figure 9.11 A chemical cell made by joining copper and zinc half-cells

Questions

15 In a Zn/Cu cell, electrons flow from the zinc to the copper through the wires and meter. Use the Data Booklet to give the direction of electron flow in each of the following cells:
a) Mg/Fe
b) Ag/Sn
c) Ni/Zn
d) Al/Pb.

16 For each of the following, state which cell would give the larger voltage:
a) Ni/Pb and Fe/Cu
b) Fe/Ag and Sn/Ag
c) Al/Hg and Mg/Au
d) Zn/Pb and Al/Hg.

Activities

This activity involves a study of 'fruity cells' that use a fruit and two different metals to produce an electric current. There are two parts to this investigation, but you will only be asked to work on one.

1 Find out which fruit makes the best electrochemical cell. You need to plan, carry out and then write a brief report on an experiment to find which fruit gives the highest voltage when it is used in a cell.

2 Find out which pair of metals (from a selection provided by your teacher) produces the highest voltage when used to make a fruity cell. You will need to plan, carry out and write up your experiment.

Decide whether you are going to do Experiment 1 or 2 then read the following advice before starting the task.

First, write down an aim for your experiment; this is what you are trying to do.

Think about the variables involved in the reaction, then write a plan to show what you intend to do and decide how you are going to record the observations.

It is usually a good idea to draw a diagram (in pencil) of the apparatus you are going to use and to make a list of the chemicals and apparatus you will need. Think about how you are going to ensure your experiments are safe. Ask your teacher/lecturer to check your plan.

When you have finished the practical work, record your results and write a 'Conclusion' (what you have found out). Always try to make the conclusion relate to the aim of your experiment.

Think about ways in which you could make the experiment more accurate. If you were going to do the experiment again, would you have done everything in the same order? Did you make any mistakes? Thinking about the answers to these questions, write one or two sentences to evaluate your experimental procedures.

Electrochemical cells as redox reactions

All electrochemical cells involve the movement of electrons from one half-cell to the other; this means that all electrochemical cells involve redox reactions. Cells are portable devices that convert chemical energy into electrical energy and in the past 50 years, a lot of work has been put into improving their design and efficiency. In many modern cells there are no metal elements involved in the redox reaction. A cell can be constructed from any chemical reaction that involves oxidation and reduction, for instance, the reaction between iron(III) chloride solution and potassium iodide solution (see Figure 9.12). The two solutions can be connected using an ion bridge as before, but the electrodes can be graphite – a conducting non-metal. Electrons flow from the iodide ions in one solution through the meter to the

iron(III) ions in the other. As this happens, the iodide ions are oxidised and turn into iodine molecules:

$$2I^- \longrightarrow I_2 + 2e^-\ [O]$$

The iron(III) ions gain electrons (are reduced) and turn into iron(II) ions:

$$(\text{yellow})\ Fe^{3+} + e^- \longrightarrow Fe^{2+}\ (\text{green})\ [R]$$

Balancing the electrons lost and gained (by multiplying the reduction reaction by two) gives:

$$2Fe^{3+} + 2e^- \longrightarrow 2Fe^{2+}$$

This equation can now be added to the oxidation equation (remembering to omit the electrons) to give the balanced redox equation:

$$2Fe^{3+} + 2I^- \longrightarrow 2Fe^{2+} + I_2$$

The spectator ions in this reaction are chloride ions and potassium ions.

Electrodes

If we make a cell by joining two different metals together, then the metals are used to conduct electricity to and from the external circuit; in other words the metals are used as electrodes. Figure 9.12 shows the cell referred to above in which neither of the electrodes is a metal. Electrons flow from one solution to the other, and so the electrodes can be any material that conducts electricity and does not react with the solution in which it has been placed. Graphite is often used for this purpose.

The next step?

Normally heart pacemakers need new batteries every seven years and that requires chest-opening surgery. Aerospace engineers in the US have developed a prototype made from magnets and a **ceramic piezoelectric** material

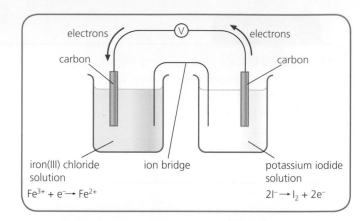

Figure 9.12 An iodide/iron(III) cell

which converts vibrations in the chest cavity (heartbeats) into electricity; the engineers have successfully used similar devices to make small-scale energy 'harvesters' for space ships. In recent years, live testing has been carried out across the world using new piezoelectric materials to try to develop energy harvesters that will produce enough electricity to power a pacemaker.

Checklist for Revision

- I understand and can explain metallic bonding theory.
- I can explain how an electric current flows in elements and compounds.
- I understand some typical reactions of metals and can explain how they occur in terms of electron flow, redox reactions, oxidation and reduction.
- I know how metals are extracted from ores by reduction reactions.
- I can write balanced ionic equations for the reactions of metals and the extraction of metals from ores.
- I can explain how electrochemical cells work, including those that utilise non-metal electrodes.

End-of-chapter questions

1 Copper can be obtained by heating copper(II) oxide with charcoal (a form of carbon) or with the gas, carbon monoxide.
   a) Write a balanced symbol equation for the reaction between:
      i) copper(II) oxide and carbon;
      ii) copper(II) oxide and carbon monoxide.
   b) Suggest why magnesium cannot be obtained from its oxide by heating with carbon.
   c) Name a metal that can be obtained from its oxide by heating alone.
2 The salt zinc(II) sulfate can be prepared by reacting zinc with dilute sulfuric acid.
   a) Describe fully how you would carry out this experiment.

   b) Write a balanced equation for the reaction.
   c) Write an ionic equation, including state symbols, for the reaction.
3* A student set up the experiment shown in Figure 9.13.

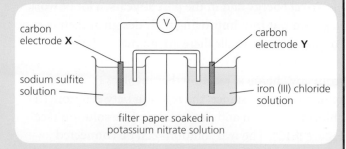

Figure 9.13

The reaction occurring at electrode Y is:

$$Fe^{3+}(aq) + e^- \longrightarrow Fe^{2+}(aq)$$

a) Using the letters X and Y indicate the direction of electron flow.

b) Name the type of reaction taking place at electrode Y.

c) The reaction occurring at electrode X is:

$$SO_3^{2-}(aq) + H_2O(l) \longrightarrow SO_4^{2-}(aq) + 2H^+(aq) + 2e^-$$

Use the ion–electron equations for the two electrode reactions to obtain a balanced redox equation for the overall cell reaction.

4 The formula for aluminium oxide is Al_2O_3.

a) What is the charge on the aluminium ion? (You may wish to use the data book to help you.)

b) The aluminium oxide can be reduced to give aluminium metal. Write an ion–electron equation to show aluminium ions changing to aluminium atoms.

5* Figure 9.14 shows the electrolysis of molten lead(II) iodide. Which of the following describes the reaction at the positive electrode?

A I^- reduced

B I^- oxidised

C Pb^{2+} reduced

D Pb^{2+} oxidised.

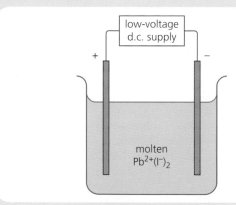

Figure 9.14

6* The electrolysis of molten sodium hydride (Figure 9.15) produces hydrogen gas.
The ion–electron equations for the reactions taking place at the electrodes are:

$$Na^+ + e^- \longrightarrow Na$$

$$2H^- \longrightarrow H_2 + 2e^-$$

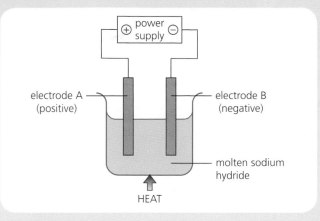

Figure 9.15

a) Use the equations to identify the electrode at which hydrogen gas is produced.

b) Combine the two ion–electron equations to give a balanced redox equation.

c) Explain how a metal, like sodium, can conduct electricity.

7* The following statements relate to four different metals, P, Q, R and S.

● In a cell, electrons flow from metal P to metal Q.

● Metal S is the only metal that reacts with cold water.

● Metal R is the only metal that can be obtained from its ore by heat alone.

The order of reactivity of the metals, starting with the *most* reactive, is:

A S, P, Q, R

B R, Q, P, S

C R, P, Q, S

D S, Q, P, R.

8* Zinc reacts with copper(II) sulfate solution.

$$Zn(s) + Cu^{2+}(aq) + SO_4^{2-}(aq) \longrightarrow$$
$$Zn^{2+}(aq) + SO_4^{2-}(aq) + Cu(s)$$

a) Rewrite the equation omitting the spectator ions.

b) When iron rusts, iron atoms are converted into iron(II) ions.

i) Write the ion–electron equation for this reaction.

ii) What name is given to this type of reaction?

10 Plastics

Worldwide, the plastics industry is worth trillions of pounds because plastics are immensely useful. In their many different forms they have greatly improved our quality of life because they are strong, lightweight, flexible, good heat and electrical insulators and cheap to produce. They do, however, have one big disadvantage; they outlive their usefulness. Plastics are synthetic (made in factories not in nature), so bacteria in the soil cannot break them down and the plastic we throw away can end up damaging the environment or clogging up a landfill site.

Figure 10.1 Toy building bricks

Activities

Many people think that it is very important to protect our environment and two good ways to do this are:

- remove plastic waste from the world's oceans
- reduce the amount of packaging used for everyday items like food and toys.

Working in a group, choose one of these ideas and use the internet to help you write a short piece about how you could plan and carry out your aim. Present your plan to the class. After each group has presented, have a class vote and decide who has developed the best scheme.

Polymers

Most people would say that a toy building brick such as Lego is made of plastic (Figure 10.1). Someone with a little knowledge of chemistry, however, would say that it is made of a **polymer**. Polymers are very big molecules made from (usually) very small molecules called **monomers**. Some polymers, like the widely used polyethylene terephthalate (PET) soften (or melt) when they are heated and these polymers can be recycled. PET is used to make most of the fizzy juice bottles that are bought each day (Figure 10.2) so recycling these bottles, and any other plastic we have used, will help reduce damage to our environment.

Figure 10.2 These fizzy drink bottles can be recycled to make new plastic products

Activities

Use the internet to find out about recycling codes for different plastics, then design and make a poster to display this information.

Look at three different plastic items at home or in school. Look for the recycling code on the item and try to find out the name of the polymer it was made from.

Alkenes

In Chapter 6, you learned that the unsaturated hydrocarbons known as alkenes can take part in addition reactions. In an addition reaction, one of the bonds in the carbon-to-carbon double bond breaks open and other atoms or groups of atoms add on to the carbon atoms in the bond. An example of an addition reaction is shown in Figure 10.3.

$$C_2H_4 + H_2 \rightarrow C_2H_6$$
$$\text{ethene} + \text{unsaturated} \rightarrow \text{ethane saturated}$$

Figure 10.3 An addition reaction

Ethene molecules can also take part in addition reactions with other ethene molecules to give a very big molecule called poly(ethene); this sort of reaction is called **addition polymerisation**.

Alkenes are one of the main **feedstocks** for the plastics (polymer) industry and poly(ethene) is the most widely used plastic in the world. Several different types of poly(ethene) have now been developed and around 80 million tonnes of the polymer are sold each year. In industry, poly(ethene) is usually called polyethylene and it is sold under the trade name of polythene. Polythene was discovered by UK chemists working in the 1930s and was one of the first synthetic polymers to be manufactured on a large scale.

Addition polymers

The addition polymers we will meet in this course are made from only one type of monomer (Table 10.1). For example, poly(propene) is made from a monomer called propene and poly(phenylethene) is made from a monomer called phenylethene. The monomers used in addition polymerisation all contain the same functional group, a carbon-to-carbon double bond, so the chemical name of each monomer ends in the letters -ene. Naming addition polymers is straightforward; the name of the monomer is put in brackets and the prefix 'poly' is placed in front of it, as shown in Table 10.1.

Figure 10.4 This polymer spinnaker sail is strong and lightweight

Monomer name	Polymer name
ethene	poly(ethene)
propene	poly(propene)
chloroethene	poly(chloroethene)
tetrafluoroethene	poly(tetrafluoroethene)
phenylethene	poly(phenylethene)

Table 10.1 Some common addition polymers

Repeating units

One thing all types of polymer molecules have in common is that they are all repetitive; the big polymer molecules are made from thousands of small monomer molecules combined together so the long chains are made of identical sections that repeat themselves over and over again. These sections are called repeating units.

Shapes of molecules

Making molecular models of compounds containing carbon shows that these molecules are rarely flat. Another term, often used instead of flat, is two-dimensional. If you put your hand flat on the desk, it is not two-dimensional because it is long, it is wide and it is also has a measurable depth or 'height'. A piece of paper placed on the desk is a better approximation of a two-dimensional object, but it still has a tiny, but measureable, height. If you look at a model of a water molecule, you will see that it is V-shaped and that it lies flat on the desk. Remember that an actual water molecule is around one million times smaller than the model and you will see why we describe water molecules as two-dimensional. By contrast, if you look at a methane molecule, you will see that no matter how many times you turn it over, it will never lie flat on the desk. Carbon atoms form four covalent bonds, so each carbon atom in methane is surrounded by four pairs of electrons, all negatively charged and all trying to get as far away from each other as possible. This electronic repulsion gives a methane molecule a particular shape: it is said to be tetrahedral. (Look back at Chapter 3 of this book to refresh your memory on shapes of molecules.)

Activities

Working in a group, make a model of a methane molecule. Stand it on a piece of paper, draw straight lines joining each of the three atoms then cut out the shape you have drawn. Turn the molecule over and repeat this process and keep doing it until you have a piece of paper for each side of the molecule. Use glue to attach a piece of paper to each 'face' of the model. How many faces, or sides, has the model got? Did you need to keep turning the model over to mark out the shapes? Use the internet to find out what the word 'tetrahedral' means.

Now make a model of an ethene molecule and see how the carbon-to-carbon double bond has forced the molecule to lie more or less flat on the desk with the hydrogen atoms pointing away from the double bond. Figure 10.5 gives the structural formula of ethene to help you.

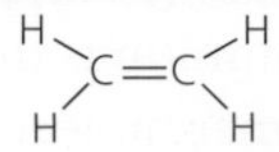

Figure 10.5 Ethene

Some common addition polymers

Poly(ethene)

Poly(ethene) (old name polyethylene, trade name polythene) is made from the monomer ethene. The structural formula for ethene is as shown in Figure 10.5; however, when we are representing monomers for use in addition polymerisation, it is best to use a 'rugby goalpost' type of structure because this focuses attention on the most important part of the molecule, the double bond. The three ethene molecules shown in Figure 10.6 are drawn in this format.

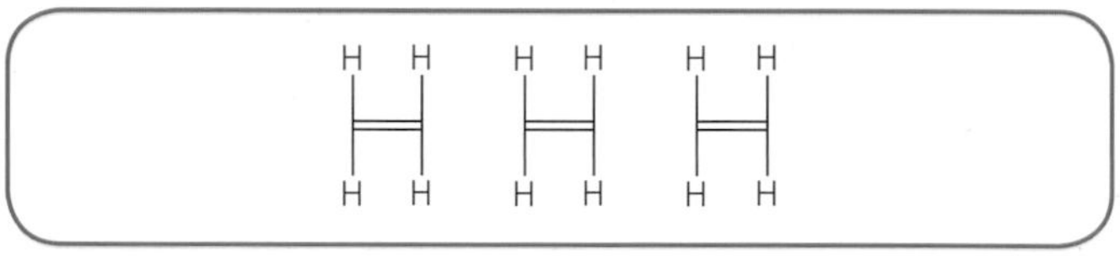

Figure 10.6 The 'rugby goalpost' structure of ethene

When the monomers are mixed in the presence of a catalyst, one of the covalent bonds in the C=C bond breaks; this means that each carbon atom is now only forming three bonds and there is a free electron on each of the carbon atoms (see Figure 10.7). These can combine with free electrons in other molecules and a chain of monomer molecules can join together.

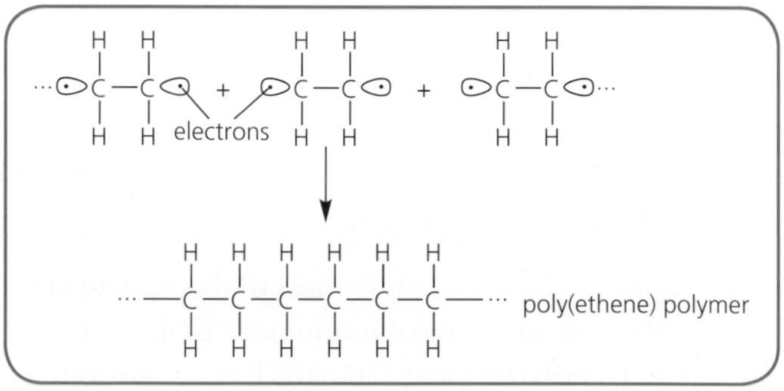

Figure 10.7 Stages in the polymerisation of ethene

All addition polymers are formed in a similar way with thousands of monomer units linked together. The structure of each repeating unit in the chain can be obtained from the structure of the monomer molecule by breaking the double bond and leaving each carbon

atom with one open bond to show it is part of a chain. For example, the repeating unit in a poly(ethene) molecule is as shown in Figure 10.8.

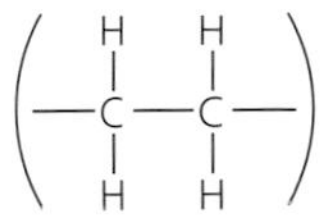

Figure 10.8 The repeating unit in a poly(ethene) molecule

The number of monomer units in a polythene molecule depends on the type of polythene that is being manufactured, so if we use the letter 'n' to mean any big number (more than a thousand), we can use the repeating unit to construct an equation for addition polymerisation (Figure 10.9).

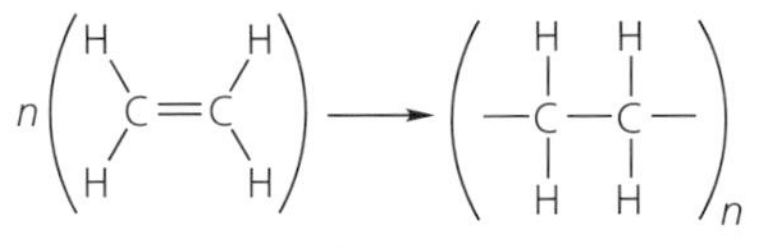

Figure 10.9 Equation for addition polymerisation

This equation can be made simpler by using molecular formulae, for example

$$nC_2H_4 \longrightarrow (C_2H_4)_n$$

(n monomer molecules)   (repeating unit repeated n times in the polymer chain)

Working in a group, use molecular models to make as many molecules of ethene as possible. Break one of the double bonds in each molecule and add all the molecules in the class together to form a polymer molecule. How many repeating units are in the molecule you have made? Discuss why the chain of carbon atoms is a zigzag shape rather than a straight line.

Activities

Poly(phenylethene)

Poly(phenylethene) (old name, polystyrene) is a **thermoplastic** polymer used to make vending cups. Blowing carbon dioxide or nitrogen gas into the polymer makes expanded polystyrene which is widely used for packaging. The molecular formula for phenylethene is C_8H_8 and the structural formula is $CH_2{=}CH{-}C_6H_5$.

The group of atoms shown as C_6H_5 is the **phenyl group**. It has a ring structure that chemists show as a circle inside a hexagon. The structure of phenylethene is shown in Figure 10.10.

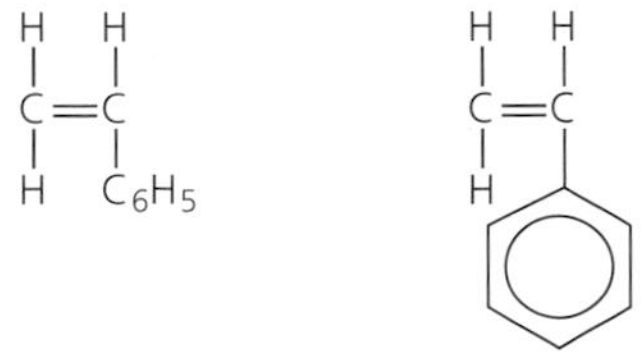

Figure 10.10 The structure of phenylethene

Looking at these 'goalpost' structures, we can see that when addition polymerisation takes place, the carbon atoms from the double bond in each monomer will combine together to form the backbone chain, as in poly(ethene) (Figure 10.11). In this instance, however, every second carbon atom in the chain is bonded to a $-C_6H_5$ (phenyl) group. It is the presence of these phenyl groups which gives poly(phenylethene) its different properties.

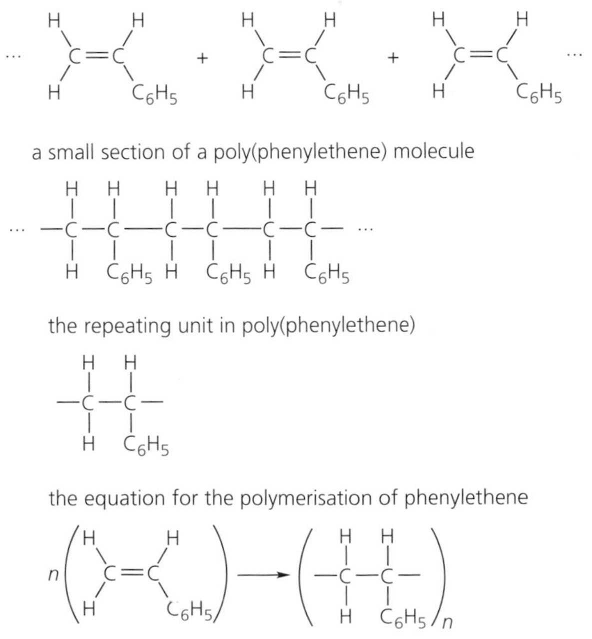

Figure 10.11 The polymerisation of phenylethene

Using molecular formulae, the equation for the polymerisation of phenylethene becomes:

$$nC_8H_8 \longrightarrow (C_8H_8)_n$$

Poly(chloroethene)

The monomer phenylethene is said to be a substituted alkene because one of the hydrogen atoms in the alkene (ethene) has been replaced by a phenyl group. This substituted alkene can still be called a hydrocarbon because it contains only hydrogen and carbon atoms but many substituted alkenes contain entirely different types of atoms. An example of this is the monomer chloroethene, the structure of which is shown in Figure 10.12.

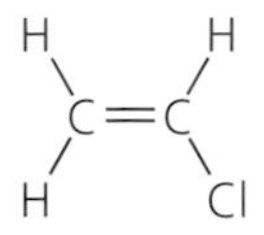

Figure 10.12 Chloroethene (a substituted ethene)

Poly(chloroethene) has a wide range of uses, mainly because substances called 'plasticisers' can be added to make the rigid polymer increasingly flexible. The old name for the monomer is vinyl chloride and the old name for the polymer is polyvinyl chloride, PVC. The rigid form of PVC is widely used to make window frames and the term 'vinyl' is generally used to describe the more flexible forms that are made into records, wallpapers and floor coverings.

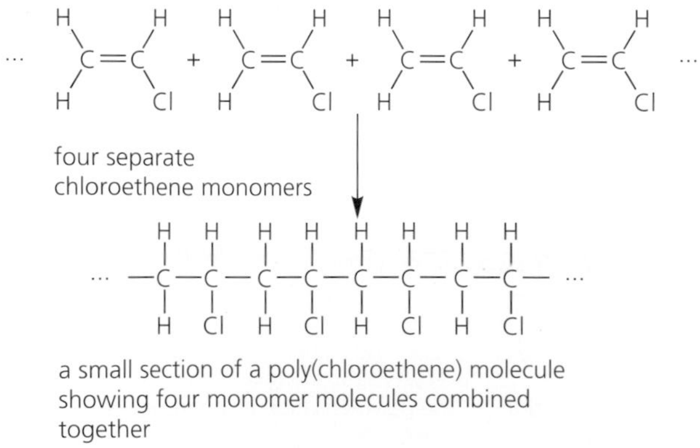

four separate chloroethene monomers

a small section of a poly(chloroethene) molecule showing four monomer molecules combined together

Figure 10.13 Polymerisation of chloroethene

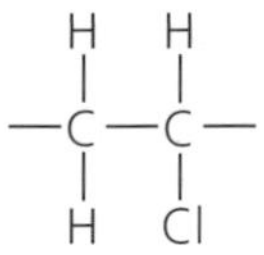

Figure 10.14 The repeating unit in poly(chloroethene)

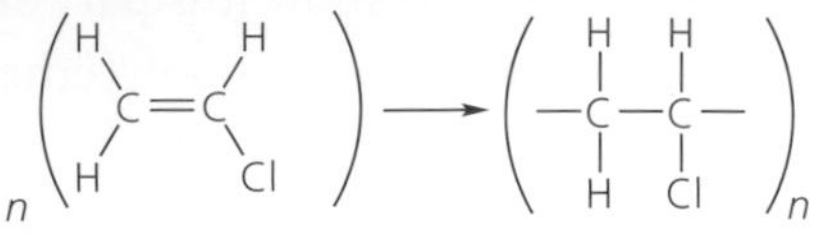

Figure 10.15 Equation for the polymerisation of chloroethene

Using molecular formulae, the equation for the polymerisation of chloroethene becomes:

$$n\,C_2H_3Cl \rightarrow (C_2H_3Cl)n$$

Activities

The information on pages 98–100 gives a detailed look at three of the five polymers listed in Table 10.1. Working in a group, choose one of the remaining polymers and use the information above and the internet to find answers to the following questions about your polymer:

- What are its main uses?
- Does it have an old name or a trade name?
- What is the structure of the monomer?
- Draw a section of the polymer showing four monomer units joined together.
- Work out the repeating unit and write equations for the addition polymerisation using first structural formulae and then molecular formulae.

Teach a 5-minute lesson to the class making sure they have got good notes on all the points you have researched. Make sure you are prepared to answer any questions they might ask about your polymer.

Checklist for Revision

- I understand how addition polymers are formed.
- I can represent the structures of monomers, polymers and polymer repeating units using structural formulae.

Activities

In this experiment, you will be making rubber in the laboratory and testing its properties.

Firstly, to make a rubber ball, place $10\,cm^3$ of latex solution in a small beaker. Add approximately $5\,cm^3$ of vinegar or citric acid solution and stir. Cross-links are formed in the mixture and a rubber ball is formed very quickly. Wearing vinyl protective gloves and eye protection, remove the rubber ball from the beaker and place it in a bowl of water. Squeeze the ball well to remove excess latex and then dry the ball with a cloth.

Secondly, repeat the above experiment, but add one spatula of sodium hydrogencarbonate to the beaker before adding the vinegar or citric acid solution to make foam rubber. Bubbles of carbon dioxide permeate the rubber ball as it forms.

Finally, to make an elastic band, place a pencil or the end of a test tube in the latex solution so that the end is well covered and then dip the end in a beaker of vinegar, followed by a beaker of water. The rubber can then be rolled off the end as a ring.

Working in a group, plan and design the following investigations:

a) A bounce height experiment to compare the elastic properties of the two rubber balls you have made.
b) An experiment to find out whether rubber is a thermoplastic or a thermosetting polymer.
c) An experiment to find out whether a long rubber band stretches more than a short one.

Make a list of the apparatus you will need and discuss your plans with your teacher/lecturer. Think about ways in which you could make your experiments more accurate.

Discuss among yourselves the order in which you will have to do these experiments, assign tasks to each member of the group, make samples of rubber, as described above, and test the samples you have made.

As a supplementary task, think about the following. You used an acid to make the liquid latex into a solid ball. Make a prediction about the type of chemical that might reverse this reaction and carry out an experiment to check your prediction.

End-of-chapter questions

1 Name the monomer used to make the addition polymer called poly(buta-1,3-diene).

2 a) Explain the meaning of the term 'addition polymerisation'.
   b) The structure of a widely used addition polymer is shown in Figure 10.16. Identify the repeating unit for this polymer and draw its full structural formula.

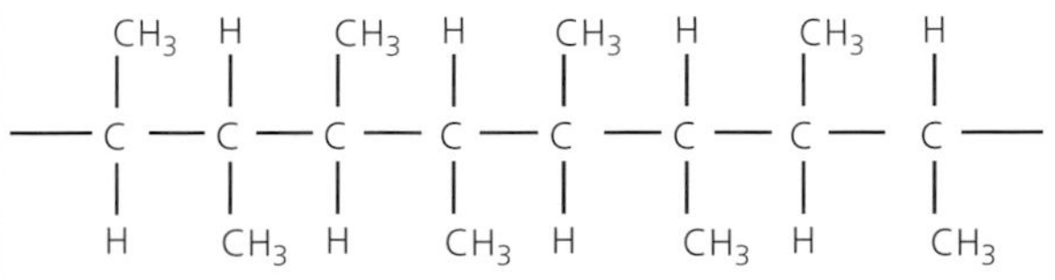

Figure 10.16

c) Draw the full structural formula for the monomer used to make this polymer.
   d) What is the systematic name for this monomer?

3 Perspex, a lightweight plastic replacement for glass, is made from the monomer shown in Figure 10.17.

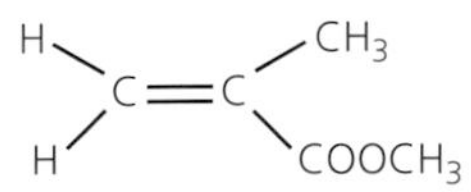

Figure 10.17

Draw
a) the repeating unit for the polymer
b) a part of the Perspex polymer chain showing three repeating units.

4 Polymer P is a versatile plastic that can be made into a film for packaging, fibres for carpets and the 'synthetic grass' used for some hockey and football pitches. Its structure is shown in Figure 10.18.

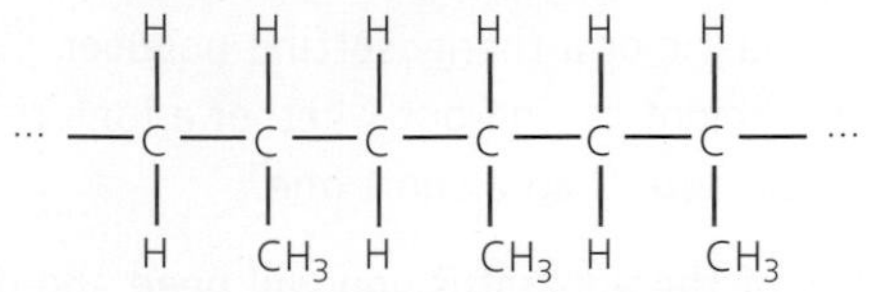

Figure 10.18

The repeating unit in the polymer is shown in Figure 10.19.

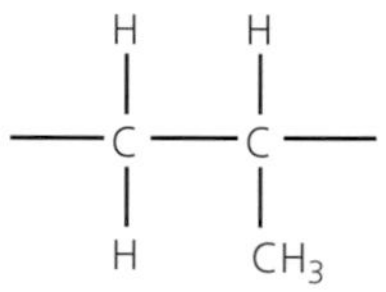

Figure 10.19

The monomer M from which the polymer is made is shown in Figure 10.20.

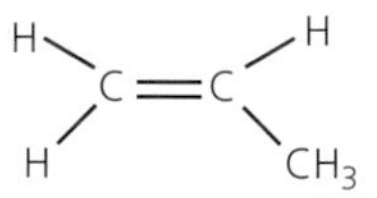

Figure 10.20

a) Name the monomer M and the polymer P.
b) To which homologous series does M belong?

5 When superglue sets, a polymer is formed. The polymer has the structure shown in Figure 10.21.

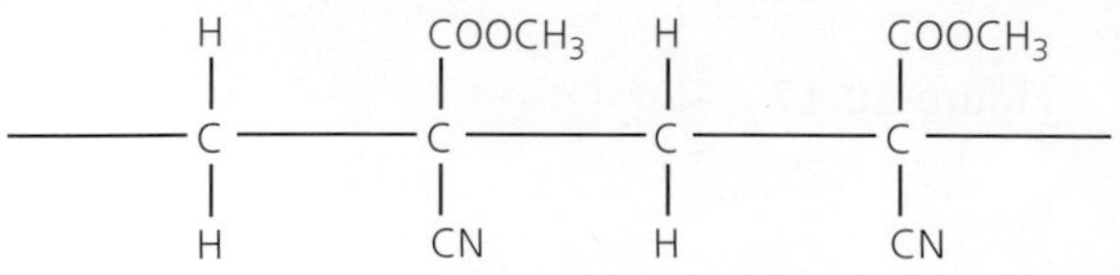

Figure 10.21

Draw the structural formula for the monomer in superglue.

6 Polyvinylidene chloride is an addition polymer. It is added to carpet fabrics to reduce flammability. It is made from the monomer vinylidene chloride. Part of the polymer molecule is shown in Figure 10.22.

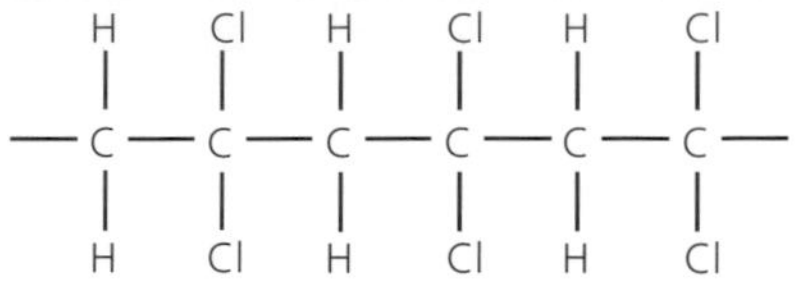

Figure 10.22

a) What is meant by an *addition* polymer?
b) Draw the full structural formula of the vinylidene chloride monomer.

7* PVC is a very useful plastic. It can be used to make window frames and to make the insulating cover for electrical cables. A short section of the polymer is shown in Figure 10.23.

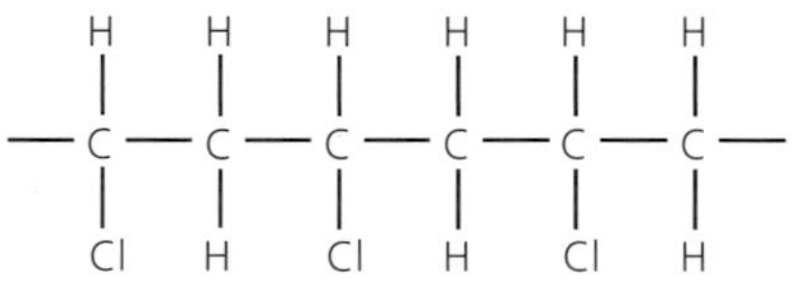

Figure 10.23

a) Draw the repeating unit present in the polymer.
b) Draw the full structural formula of the monomer used to produce the polymer PVC.

8* Polymers are used in the manufacture of glass windscreens (Figure 10.24). This makes the windscreen less likely to shatter into fragments.

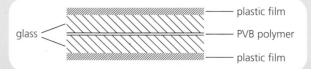

Figure 10.24

The PVB polymer is made from a monomer with the structure shown in Figure 10.25.
Draw a section of the polymer showing three monomer units linked together.

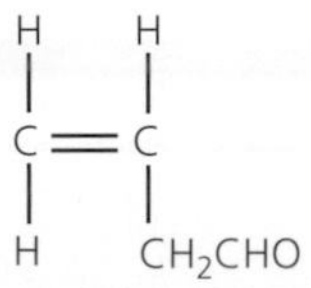

Figure 10.25

9* Poly(ethenyl ethanoate) is an addition polymer. Part of its structure is shown in Figure 10.26.

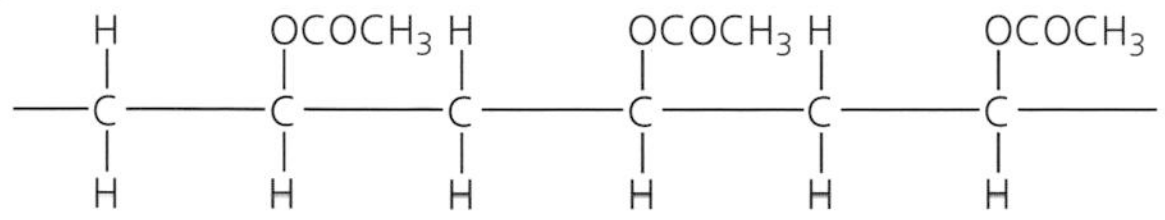

Figure 10.26

Draw the structural formula for the monomer used to make this polymer.

10 A non-stick frying pan is made of metal coated with poly(tetrafluoroethene). A section of the polymer is shown in Figure 10.27.

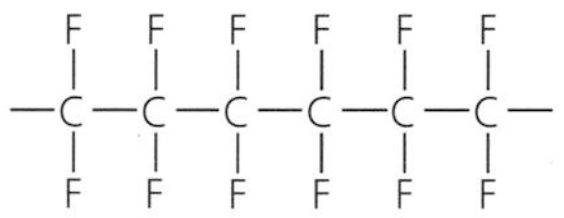

Figure 10.27

a) What type of polymer is poly(tetrafluoroethene)?

b) Draw the structure of the monomer from which this polymer is made.

11 Some audio equipment contains a material called poly(ethyne). Part of the structure of poly(ethyne) is shown in Figure 10.28.

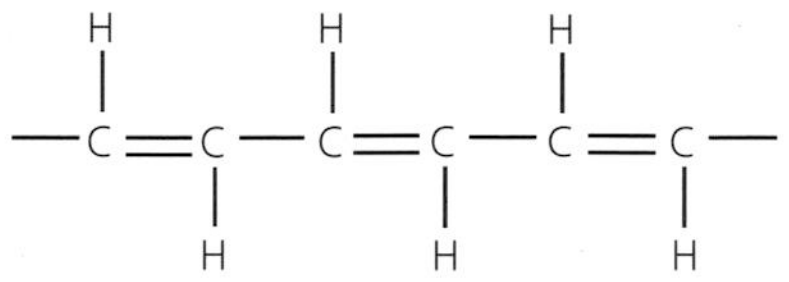

Figure 10.28

The monomer for this polymer belongs to a homologous series called the alkynes.

a) Draw the full structural formula for the repeating unit present in the polymer.

b) Draw the structural formula for ethyne.

11 Fertilisers

Food security

Food security is a term frequently used in news programmes. The UN Food and Agricultural Organization defines food security as, 'access to sufficient, safe and nutritious food to meet [a person's] dietary needs and food preferences for an active and healthy life'. In this chapter, we will look at how twentieth century chemists developed the artificial fertilisers that help feed the world and provide food security for millions of people. Synthetic and natural fertilisers have had a great impact on the world's food production. The importance of fertiliser manufacture can be seen if we consider a few facts and figures.

World population

A look at the US census site (given below) will show you an estimate of the number of people who presently share this planet.

www.census.gov/main/www/popclock.html

The 'Green Revolution' of the twentieth century increased food supplies and population growth followed. Experts agree that the world's population is still increasing (although the rate of growth is now decreasing) and the statistics indicate that there will be over 9 billion of us by 2050. Artificial fertilisers play a major role in feeding hungry people, but their manufacture depends on the availability of fossil fuels, and fossil fuels are a finite resource. Water is also becoming a vital resource (Figure 11.1), as patterns of rainfall change across the planet, so new methods of irrigation and water storage need to be developed.

Elements essential for plant growth

Healthy crops provide more food and to be healthy, plants need about 16 different chemicals, the three most important ones being nitrogen, phosphorus and potassium (NPK). When crops grown locally were eaten locally, these elements could be returned to the soil by using natural fertilisers. As more people moved into towns, crops were harvested and, to feed the townspeople, they were taken miles from where they had been grown. As a result, natural fertilisers were no longer enough to nourish the soil. Nitrogen is one of the most important elements needed by plants and for many years, ships ferried guano (bird droppings) from Peru to Europe to use as fertiliser because guano contains a lot of nitrogen, as well as phosphorus and other useful chemicals. If you go to the Isle of May in summer, when the sea birds are nesting, you can see guano on the cliffs (Figure 11.2). In this country, it does not stay on the cliffs for long because it is soluble in water, another property that a fertiliser must possess, and so gets washed away by the rain. In the dry conditions of South America, huge deposits built up and, during the nineteenth century, they were dug out and used to fertilise the fields of Europe.

Figure 11.1 The Sahara desert

Figure 11.2 Sea cliffs white with guano

Nitrogen fixation

Farmers have known for possibly thousands of years that some leguminous plants, such as peas, beans and clover, can be used as 'green compost'. In other words, they can be grown and then ploughed straight back into the soil to improve its fertility. This is because these plants contain colonies of bacteria which trap (or fix) nitrogen from the air and use it for growth. Although it makes up 80% of our atmosphere, nitrogen is an unreactive gas; it is difficult to fix. A brilliant chemist called Fritz Haber started working on this problem in 1900. By 1908 he had successfully fixed atmospheric nitrogen in the laboratory to make the compound called ammonia (NH_3) and the stage was set for one of the biggest developments of the twentieth century: intensive agriculture.

Formation of ammonia from its elements

There is a more or less limitless supply of nitrogen in our atmosphere and hydrogen can be obtained from natural gas (methane), so the elements used to make ammonia are relatively cheap and easily available. The compound itself was first isolated in the eighteenth century, but back then it was not prepared from its elements, and the method of preparation made it too expensive to be produced industrially. Haber's work inspired chemists and engineers to scale the process up from a laboratory reaction to a profitable chemical industry. To make money, the industry needed a fast reaction that gave a high yield of ammonia at the lowest possible cost, but scientists faced several problems.

1 The reaction between nitrogen (N_2) and hydrogen (H_2) to make ammonia (NH_3) is reversible:

$$N_2(g) + 3H_2(g) \rightleftharpoons 2NH_3(g)$$
$$\text{(reactants)} \qquad \text{(product)}$$

The double arrow indicates that the reaction is reversible. Reversible means that nitrogen and hydrogen are combining to form ammonia, but, at the same time,

ammonia is breaking down to form nitrogen and hydrogen.

When N_2 and H_2 are first mixed and start to react, the concentration of the elements is high and the concentration of the compound (NH_3) is low (because very little of it has yet formed). As you know from Chapter 1 of this course, a high concentration of reactants means a fast reaction, and vice versa, so initially the reaction from left to right is fast and the reaction from right to left is slow. As more of the N_2 and H_2 react and get used up, the reaction from left to right (usually called the forward reaction) slows down and, as more of the NH_3 is formed, the reaction from right to left (usually called the reverse reaction) speeds up. Eventually, there comes a point at which ammonia is breaking down at the same rate as it is being formed. This means that it is impossible to change all the nitrogen and hydrogen into ammonia. In fact, this reaction gives a low yield of ammonia.

2 The reaction is also slow at room temperature. Both forward and reverse reactions can be speeded up by increasing the temperature, but ammonia decomposes much faster at higher temperatures, so increasing the temperature decreases the ammonia yield still further.

3 The reactants (N_2 and H_2) are gases, so the molecules are far apart, yet the molecules must collide if they are to react. Exerting pressure on the system 'squeezes' the molecules closer together and increases the yield of ammonia, but running an industrial process at high pressures is costly and the higher the pressure, the greater the production costs.

It took 5 years for Haber, working with a research team led by the engineer Karl Bosch, to develop the first industrial-scale application of the Haber process (Figure 11.3). The process was carried out in a sealed system kept at high pressure. A catalyst (powdered iron) was used to get a fast reaction at a relatively low temperature. The difference in boiling points between the three gases, N_2, H_2 and NH_3, allowed the engineers to condense ammonia and run it off as a liquid, leaving the unreacted gases (nitrogen and hydrogen) to be recycled to avoid wasting reactants.

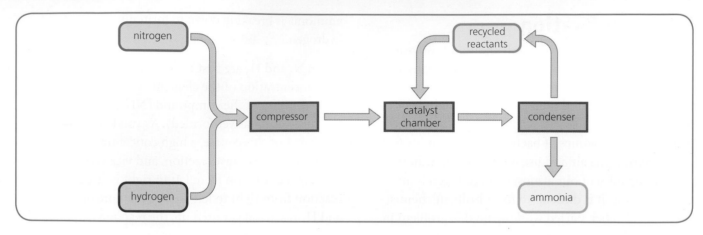

Figure 11.3 In the Haber process, nitrogen and hydrogen are combined to make ammonia

Ammonia

Ammonia is an important chemical because it is a feedstock for the manufacture of nitric acid (Figure 11.4) and for the fertiliser industry.

Properties of ammonia

Ammonia gas consists of NH_3 molecules. It is soluble in water because the molecules react with water to form ammonium ions, $NH_4^+(aq)$, and hydroxide ions, $OH^-(aq)$. This is a reversible reaction.

Figure 11.4 An industrial nitric acid production plant

Activities

Industrial chemists in different countries use different conditions for the production of ammonia, depending on the plant they are operating. Go to the website **www.freezeray.com/chemistry.htm** and use the animation of the Haber process to find out how the yield of ammonia changes as the conditions of temperature and pressure change.

Construct a graph with percentage yield of ammonia on the vertical axis and pressure in atmospheres (atm) on the horizontal axis. Set the temperature at 200 °C and the pressure at 100 atm, discharge the ammonia as soon as 'equilibrium' is reached and note the yield. Keep the temperature constant, increase the pressure to 200 atm and note the yield again. Note the yield of ammonia at 400 atm, 600 atm, 800 atm and 1000 atm, then plot the readings on the graph. Draw a line of best fit and label the line 200 °C.

Choose two different temperatures and repeat the process for each temperature. You now have three lines on your graph. Decide whether these three give you enough information about yields in the Haber process. Should you draw a fourth?

Remembering that high pressures generate high running costs, use your graph to estimate the most cost-effective conditions for the manufacture of ammonia. Use the internet to check your prediction.

Be careful not to get ammonia molecules, NH_3, confused with ammonium ions, NH_4^+.

$$NH_3(g) + H_2O(l) \rightleftharpoons NH_4^+(aq) + OH^-(aq)$$

Ammonia solution (sometimes called ammonium hydroxide solution) has a pH above 7 and turns universal indicator solution blue. Ammonia is the only alkaline gas you will meet in this course – in other words, the only gas that turns damp pH paper blue. It does this because it dissolves in the water in the damp paper to form hydroxide ions.

The high solubility of ammonia can be shown by the fountain experiment (Figure 11.5).

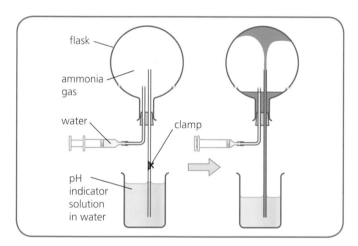

Figure 11.5 All the ammonia gas in the flask can dissolve in just a few drops of water. When this happens, the water from the beaker rushes into the flask to fill the space, creating a fountain effect as soon as the clamp is released

Care should be taken when using ammonia gas. It is the main active ingredient of smelling salts and it irritates the mucous membranes of the nose and lungs.

Laboratory preparation of ammonia

Ammonia can be made by heating any solid ammonium compound with a solid alkali, such as soda lime or calcium hydroxide. If solutions are used instead of solids, the gas formed immediately dissolves in the water, rather than being released.

calcium hydroxide + ammonium sulfate $\longrightarrow$
calcium sulfate + ammonia + water

Activities

Plan and design an experiment to prepare and test some ammonia. Draw a diagram (in pencil) of the apparatus you will need and write down how you intend to test for the presence of ammonia gas. List the chemicals you will use, then risk assess each chemical and each procedure. Think about how you will heat the mixture of solids safely and discuss your plans with your teacher/lecturer.

If you are allowed to continue, set up the apparatus then prepare and test the gas. While the apparatus is cooling, record your observations and write down two ways in which you could have improved the experiment.

How did you know that you had made ammonia gas? How could you have collected a dry sample of the gas, if you had wanted to? (Hint: look up the density of ammonia gas.)

Check that the apparatus is cool, then dismantle it, wash and dry the glassware and return it to the correct place.

Reactions of ammonia

Reaction of ammonia with acid

In the same way that ammonia reacts with water to form ammonium ions and hydroxide ions, ammonia reacts with acids to form ammonium salts. For example, ammonia reacts with nitric acid to form the salt ammonium nitrate:

$$NH_3(g) + HNO_3(aq) \longrightarrow NH_4^+(aq) + NO_3^-(aq)$$
ammonia + nitric acid $\longrightarrow$ ammonium nitrate

In the laboratory this salt is usually prepared by adding ammonia solution (ammonium hydroxide solution) to nitric acid:

$$NH_4OH(aq) + HNO_3(aq) \longrightarrow NH_4NO_3(aq) + H_2O(l)$$
ammonium + nitric $\longrightarrow$ ammonium + water
hydroxide   acid   nitrate

107

Using ionic formulae, the above equation can be written as:

$$NH_4^+(aq) + OH^-(aq) + H^+(aq) + NO_3^-(aq) \longrightarrow$$
$$NH_4^+(aq) + NO_3^-(aq) + H_2O(l)$$

Plants can only obtain the essential elements they need for healthy growth if compounds containing those elements are dissolved in soil water, the water that plants absorb through their roots. This means that fertilisers have to be soluble in water. A look at the solubility table in your Data Booklet will tell you that all ammonium compounds are soluble in water, so as ammonium compounds also contain nitrogen, they are often used as fertilisers.

Oxidation of ammonia

Ammonia can, in the presence of a catalyst, be oxidised, and this is the first step in the manufacture of nitric acid.

Many common fertilisers are salts of nitric acid, so we shall first look at the manufacture of nitric acid before we look at the huge agrochemical industry involved in the manufacture of fertilisers.

Oxides of nitrogen

As you learned in Chapter 5 of this book, soluble non-metal oxides form acids when they dissolve in water. When nitrogen dioxide (NO_2) dissolves in water, nitric acid is formed.

Nitrogen exists in the atmosphere as diatomic molecules with the two atoms in the N_2 molecule held together by a triple covalent bond. It takes a lot of energy to break these three bonds (to separate the atoms), so atmospheric nitrogen is unreactive. If, however, enough energy is available to split the N_2 molecules, the free atoms released are very reactive. Flashes of lightning (Figure 11.6) and the spark plugs in cars provide enough electrical energy to split nitrogen molecules causing nitrogen dioxide to form. Other oxides of nitrogen also exist. Ultra-violet light (sunlight) can split up NO_2 to give NO (nitrogen monoxide) which can then recombine with oxygen to form NO_2. Nitrogen dioxide can dimerise to make dinitrogen tetroxide, and then there is N_2O which used to be used by dentists as an anaesthetic and is commonly called laughing gas.

It requires a huge amount of electrical energy to make NO_2 directly from N_2, so it is too expensive to make nitric acid commercially this way. In 1902, Wilhelm Ostwald developed (and with great forethought, patented) a method of converting ammonia into nitric acid using platinum as a catalyst. As soon as ammonia started to be mass produced, the feedstock became cheap and easily available, and the Ostwald process rapidly became a mainstay of the chemical industry. At present, some 60 million tonnes of nitric acid are produced annually, of which 85% is used to make fertilisers.

Figure 11.6 Flashes of lightning help to add nitrogen to the soil

Questions

2 You know the meaning of the word 'polymerise', so deduce the meaning of the word 'dimerise'.
3 Write the formula for the compound called dinitrogen tetroxide.
4 What is the correct name for the compound N_2O?
5 In the presence of ultra-violet light, one molecule of nitrogen dioxide and one molecule of oxygen react to produce one molecule of nitrogen monoxide and one molecule of another substance. Write the equation for this reaction and, by balancing the atoms on both sides, find the formula for the unknown substance. Use the internet to find its name and to find out about the role it plays in the atmosphere.

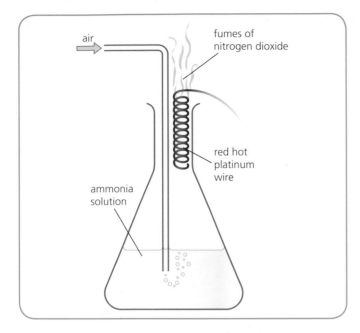

Figure 11.7 The laboratory oxidation of ammonia

Oxidising ammonia in the laboratory

Your teacher/lecturer may demonstrate the experiment shown in Figure 11.7.

The gas mixture used in the Ostwald process is approximately 10% ammonia and 90% air, so a stream of air needs to be pumped into the flask of ammonia to keep this reaction going. Heat is needed to start the reaction, so the piece of platinum wire needs to be red hot before it is placed in the flask. Once the reaction starts, the platinum continues to glow red because it is an exothermic reaction and the heat from the reaction keeps the wire hot. The ammonia is oxidised to form nitrogen monoxide and water.

As the nitrogen monoxide reaches the mouth of the flask, it combines with more oxygen to form nitrogen dioxide, which is a brown gas. If this brown gas is tested with damp pH paper, the paper will turn red as the NO_2 reacts to form nitric acid.

The Ostwald process

In the Ostwald process (Figure 11.8), the mixture of gases (ammonia and air) is heated then passed through the layers of platinum gauze that act as a catalyst. The gauze does not need to be heated separately: the oxidation of ammonia is exothermic, so the temperature stays high in the catalyst chamber and the reaction occurs at around 800 °C. The oxidation of ammonia is one of the most efficient catalytic reactions, with conversion of up to 98%.

ammonia(g) + oxygen(g) ⟶
nitrogen monoxide(g) + water(g)

The nitrogen monoxide formed reacts with oxygen to give nitrogen dioxide.

nitrogen monoxide(g) + oxygen(g) ⟶
nitrogen dioxide(g)

The brown gas is then cooled, mixed with more air and passed through a flow of water to make nitric acid.

nitrogen dioxide(g) + oxygen(g) + water(l) ⟶
nitric acid(aq)

Questions

6 Why is the platinum catalyst used in the form of a gauze?
7 Write and balance a chemical equation for each of the word equations in this section.

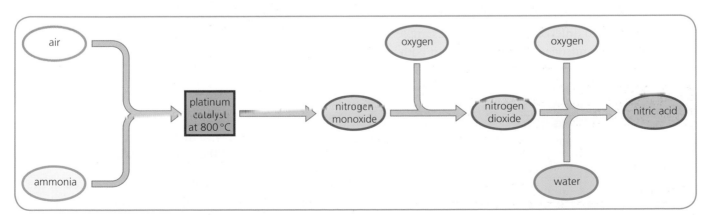

Figure 11.8 An outline of the Ostwald process

Uses of nitric acid

Nitric acid is unusual. It played a part in saving the lives of millions of hungry people across the planet through its use in the manufacture of fertilisers and it played a part in killing millions of people through its use in the manufacture of explosives. Nitric acid proves the point that there is no such thing as a dangerous chemical, there are only dangerous people.

Fertilisers

If plants are grown in the same field year after year, the soil loses its nutrients, so fertilisers are used to replace them (Table 11.1). The use of fertilisers allows farmers to increase the amount of crops they grow, and that means more food can be produced and intensive agriculture becomes possible. Farmers mainly use artificial fertilisers to supply nitrogen, phosphorus and potassium (Figure 11.9). Plants take in these nutrients with the water they absorb through their roots, so fertilisers must be soluble. The solubility table in your Data Booklet will tell you that some types of compounds are more soluble than others, which is why compounds commonly used in fertilisers include ammonium phosphate, $(NH_4)_3PO_4$, and potassium sulfate, K_2SO_4.

Ammonia and nitric acid are generally produced in the same integrated chemical plant and are then often reacted together, on site, to produce ammonium nitrate, a very soluble salt with a high nitrogen content. The big fertiliser factories are mainly located in the former USSR, the USA and in central Europe, but large quantities of potassium chloride and mineral phosphates are being produced from the Dead Sea (Figure 11.10). These were sold as raw materials, but are increasingly being converted into compound fertilisers at the site of their extraction.

Chemical companies now 'tailor' their fertilisers for different crops and different countries. For example, cereals need more nitrogen than potatoes, but both need

Figure 11.9 A bag of NPK fertiliser containing the nutrients nitrogen (N), phosphorus (P) and potassium (K)

Figure 11.10 Salt pans beside the Dead Sea

phosphorus, potassium and other minerals, and farmers in a country with a heavy rainfall might use a less-soluble compound (like urea) so that the expensive fertiliser will not be washed out of the soil before plants can absorb it.

Compounds used as fertilisers can be produced in the laboratory by neutralisation reactions similar to those you looked at in Chapter 5 of this book.

Elements needed by plants	Compounds present in fertiliser
potassium	potassium salts
phosphorus	phosphates
nitrogen	nitrates and ammonium salts

Table 11.1

Questions

8   Copy and complete Table 11.2. The first row has been done for you.

Acid	Formula of acid	Ions present in aqueous solution
phosphoric	H_3PO_4	$3H^+(aq)$ and $PO_4^{3-}(aq)$
sulfuric		
nitric		
hydrochloric		

Table 11.2

9   Phosphoric acid, H_3PO_4, is used to make phosphate fertilisers and can be prepared by the reaction of calcium phosphate with sulfuric acid. Given that calcium sulfate is the only other product, write a balanced chemical equation for the reaction.

10 Potassium chloride can be made by neutralising hydrochloric acid with potassium hydroxide.
    a) Write and balance a chemical equation for this reaction.
    b) Name the spectator ions in the reaction.
    c) Write an equation for the reaction omitting the spectator ions.

11 Write and balance chemical equations for each of the following word equations.
    a) potassium hydroxide + nitric acid → potassium nitrate + water
    b) ammonia + hydrochloric acid → ammonium chloride
    c) ammonia + phosphoric acid → ammonium phosphate

12 Complete the following word equations then write a balanced chemical equation for each of them.
    a) potassium hydroxide + sulfuric acid →
    b) sodium hydroxide + nitric acid →
    c) potassium hydroxide + phosphoric acid →
    d) ammonia solution + sulfuric acid →

The percentage mass composition of fertilisers

When fertilisers are prepared for sale, different chemical compounds are mixed in different proportions for different uses, so it is important to know the exact quantity of each element in fertilisers. To establish this we need to know how to calculate the percentage mass composition of a compound. First, we will revise some earlier work.

Formula mass

The formula mass of a compound is found by adding together the relative atomic mass of each atom in the formula. Values for the relative atomic masses of some elements can be found in the Data Booklet.

For example, if the formula is NaOH, the formula mass is $23 + 16 + 1 = 40$.

If more than one atom of an element is present in the formula, the relative atomic mass of the element is multiplied by the appropriate number.

For example, if the formula is Na_2CO_3, the formula mass is $(2 \times 23) + 12 + (3 \times 16) = 106$.

If the formula of the compound contains ions with more than one type of atom (usually called group ions), the mass of every atom inside the bracket is multiplied by the number outside the bracket.

For example, if the formula is $(NH_4)_2SO_4$, the formula mass is $(2 \times 14) + (8 \times 1) + 32 + (4 \times 16) = 132$.

Remember that in order to avoid unforced arithmetic errors, it is often safer to calculate the mass of the atoms inside the bracket then multiply that mass by the number outside the bracket.

For example, if the formula is $(NH_4)_2SO_4$, the formula mass is $[14 + (4 \times 1)] \times 2 + 32 + (4 \times 16) = 36 + 32 + 64 = 132$.

Percentage composition

Once we have calculated the formula mass of a compound, we can find the percentage by mass of any element in the compound by using the following equation:

$$\text{percentage mass of a given element in a compound} = \frac{\text{mass of element in the formula}}{\text{formula mass}} \times 100$$

Chemistry in Society

Worked examples

1 percentage of sodium $=\dfrac{\text{mass of sodium in formula}}{\text{formula mass}} \times 100$
   in sodium hydroxide

   $= \dfrac{23}{40} \times 100 = 57.5\%$

2 percentage of carbon in $= \dfrac{12}{106} \times 100 = 11.32/11.3\%$
   sodium carbonate

3 percentage of oxygen in $= \dfrac{64}{132} \times 100 = 48.48/48.5\%$
   ammonium sulfate

Checklist for Revision

- I know the best operating conditions for the Haber process to produce ammonia and can explain why they are applied.
- I know the operating conditions for the Ostwald process.
- I appreciate the need for the commercial production of nitrate fertilisers and can explain how they can be manufactured from nitric acid.

End-of-chapter questions

1 Calculate the percentage by mass of nitrogen in each of the following compounds which are used as fertilisers:
   a) sodium nitrate
   b) calcium nitrate
   c) ammonium sulfate
   d) ammonium nitrate.

2 Potassium is a major plant nutrient. It promotes the rate of plant growth. Potassium is usually added to soils in the form of the salt potassium chloride. Calculate the percentage by mass of potassium in potassium chloride.

3 Ammonia gas is converted into various solid compounds that are used as fertilisers. In one such reaction, the compound urea is formed. Molecules of urea have the structure shown in Figure 11.11.

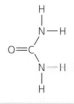

Figure 11.11

   a) i) Calculate the formula mass of urea.
      ii) Calculate the percentage by mass of nitrogen in urea.
   b) Ammonia and carbon dioxide react together, using high pressure and temperature, to give urea and water. Write a balanced equation for this reaction.

4 The chemical industry produces large quantities of nitric acid by the catalytic oxidation of ammonia (Figure 11.12).

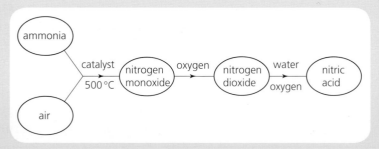

Figure 11.12

   a) State the name of this industrial process.
   b) Name the catalyst used to speed up the reaction.
   c) The nitric acid can be used to make potassium nitrate fertiliser. Write the formula for potassium nitrate.

5 As the world population increases, the demand for food grows. In order to meet this demand, farmers are using more and more synthetic fertilisers to improve crop yields. One of these synthetic fertilisers is Nitram. The flow diagram in Figure 11.13 shows how Nitram can be made industrially.
   a) i) Name reactant A.
      ii) Name industrial process X.

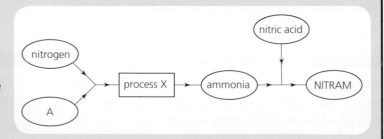

Figure 11.13

b) In process X, the percentage conversion of nitrogen to ammonia decreases as the temperature increases. Why, then, is process X carried out at the relatively high temperature of 450 °C?

c) What is the chemical name for Nitram?

d) Nitram is very soluble in water and this allows essential elements to be taken in by the roots of crop plants very quickly. Suggest why Nitram's high solubility can also be a disadvantage in its use as a fertiliser.

6 Plants use a compound called chlorophyll to obtain energy from the Sun, and nitrogen is one of the elements needed to make chlorophyll. Most plants get the nitrogen they need from compounds of ammonia and nitrates in the soil.

a) Name a substance that would produce ammonia if it was heated with ammonium carbonate.

b) The Ostwald process is used to produce nitric acid. The first stage in this process is the oxidation of ammonia.

i) Which metal is used as a catalyst in the Ostwald process?

ii) Why is it not necessary to keep heating the catalyst once the reaction has started?

7 Diammonium hydrogenphosphate is a major fertiliser made from ammonia and phosphoric acid.

$$2NH_3 + H_3PO_4 \rightarrow (NH_4)_2HPO_4$$

Calculate the percentage, by mass, of nitrogen and phosphorus in this fertiliser.

8 Phosphorus is an important plant nutrient as it regulates leaf development and size. Phosphorus is found as calcium phosphate, $Ca_3(PO_4)_2$, but this is converted into calcium dihydrogenphosphate, $Ca(H_2PO_4)_2$, which is then used as a fertiliser.

a) Calculate the percentage by mass of phosphorus in

i) calcium phosphate

ii) calcium dihydrogenphosphate.

b) Calcium phosphate is not used directly as a fertiliser, but calcium dihydrogenphosphate is. Suggest a reason for both of these facts. (The table of solubilities in the Data Booklet may help you to answer this question.)

12 Nuclear chemistry

The particle zoo

Figure 12.1 Professor Brian Cox

The range of sub-atomic particles found in the Universe is sometimes called the particle zoo. You learned about protons, neutrons and electrons in Chapter 2 of this book, but how about glueons, quarks, mesons, muons and neutrinos? If you follow Robin Ince or Brian Cox (Figure 12.1) on Twitter, then you might have heard of some of these particles. You may also have heard of the Higgs-Boson particle, which scientists at CERN in Switzerland believe they may have found. This particle is thought to convert energy into matter. At present that fact is uncertain, but it is certain that we, on this planet, can convert matter from the nucleus of an atom into energy; this is called nuclear energy.

Nuclear radiation

We live our lives in an invisible ocean of electromagnetic radiation. Radios, televisions, mobile phones (Figure 12.3), microwave ovens and WiFi networks all use electromagnetic radiation of varying wavelengths. Nuclear radiation has a different source; it is emitted from the nucleus of a radioactive isotope and it may be in the form of particles or waves.

Activities

Marie Curie and Albert Einstein were two of the scientists who revolutionised nuclear science and each faced significant barriers as they attempted to pursue their scientific work.

As you work through this section on nuclear chemistry, create a timeline following the work of these two Nobel Prize winners and the development of the understanding of nuclear radiation and its applications through the twentieth and twenty-first centuries.

Figure 12.2 At the Solvay Conference held in 1927, many of the most famous scientists of the day gathered to discuss quantum theory

Figure 12.3 This smartphone uses electromagnetic radiation

Isotopes and nuclide notation

An atom of an element with a particular mass number is called an isotope or nuclide (nuclide is the term mainly used in the nuclear industry). The existence of isotopes means that scientists use nuclide notation to show the number of sub-atomic particles in an atom or ion. As you learned in Chapter 2 of this book, isotopes are atoms that have the same atomic number (so they have the same number of protons) but they have different mass numbers (so they have different numbers of neutrons). Almost all elements have more than one naturally occurring isotope; the number varies from two up to about ten.

Hydrogen has three isotopes. The symbols for the three types of hydrogen atom are as shown in Figure 12.4.

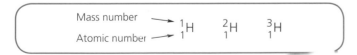

Mass number $\longrightarrow$ ^1_1H ^2_1H ^3_1H

Atomic number $\longrightarrow$

Figure 12.4 The three isotopes of hydrogen

Rather than use superscripts and symbols to give the mass number of an isotope, many textbooks use names and numbers. For example, $^{241}_{95}Am$ can be written as americium-241. The atomic number is left out because the Data Booklet will tell you that if the element is americium, the atomic number must be 95. The mass number is stated because it changes as the isotope changes.

Radioisotopes

As we move through the Periodic Table, the atomic number of the elements increases and atoms get bigger and heavier with the number of neutrons in the nucleus increasing more rapidly than the number of protons. As the ratio of neutrons to protons increases, the nuclei can become unstable. An unstable nucleus (or atom) is said to be radioactive. These radioactive nuclei contain a lot of energy and they tend to rearrange themselves spontaneously to reduce the size of the nucleus and emit some energy. This process is called nuclear decay and isotopes that decay are called **radioisotopes**. There are a relatively small number of naturally occurring radioisotopes, but there are many synthetic radioisotopes which are manufactured for use in medicine and industry.

Questions

1 The most common isotope of potassium is ^{39}K (93.3% abundance), whereas the most common isotope of uranium is ^{238}U (99.3% abundance). Work out the number of protons and neutrons in each of these isotopes.
2 Which element is commonly used as a fuel in fission reactors?

Alpha (α), beta (β) and gamma (γ) radiation

Many things can happen when a nucleus decays, but in this course, we are concerned with the emission of alpha (α), beta (β) and gamma (γ) radiation.

Alpha

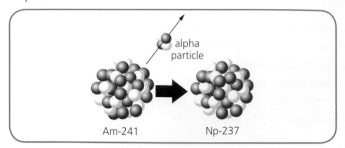

Figure 12.5 Alpha decay of Am-241 to form Np-237 plus an alpha particle

When some radioactive nuclei decay, they eject a 'chunk' of the nucleus consisting of two protons and two neutrons (Figure 12.5). This is known as an **alpha particle** and it is identical to the nucleus of a helium atom. It can be represented as ^4_2He. The particle carries a positive charge because it has two protons, but no electrons. The charge is not usually included in the symbol of an alpha particle, which can also be written as $^4_2\alpha$. Alpha particles have a mass of 4, but cannot travel very far: they can be stopped by a sheet of paper or a few centimetres of air. We say they have a low penetrating power.

Beta

Some nuclei become stable by converting a neutron into a proton and an electron:

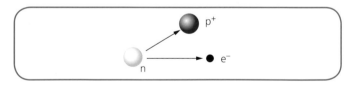

$$^1_0\text{n} \longrightarrow {}^1_1\text{p} + {}^{\,0}_{-1}\text{e}$$

(Note that a proton can also be written as ^1_1H.)

The proton stays in the nucleus, but the electron is thrown out of the nucleus at high speed and is called a **Beta (β) particle**. Beta particles are electrons produced *in the nucleus*; they are negatively charged, but they do not come from the orbiting electrons. Beta particles have negligible mass and they have greater penetrating power than alpha particles; they can pass through a sheet of paper, but are absorbed by a thin sheet of metal.

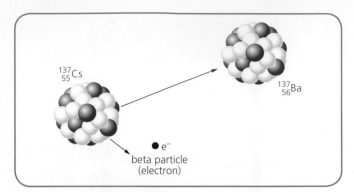

Figure 12.6 Beta decay of Cs-137 to form Ba-137 and an electron

Gamma

Gamma rays ($^0_0\gamma$) are electromagnetic waves which have an even shorter wavelength than X-rays. Gamma rays have no mass and no charge, they are not deflected by electric or magnetic fields, they can travel long distances in air and can only be stopped by thick layers of concrete (Figure 12.7) or lead. Gamma emission does not change the mass or charge of the nucleus; it just releases some excess energy.

Nuclear equations

Nuclear equations are used as a way of showing the changes that take place in the nucleus of radioisotopes; the electrons arranged outside the nucleus are ignored.

In nuclear equations, alpha or beta decay (or emission) means the particle is leaving the isotope, so the symbol for the particle emitted is shown on the right of the arrow. Capture of a particle means that the isotope has absorbed the particle, so the symbol for the particle captured is shown on the left of the arrow.

The following equation shows the alpha decay of americium-241 to form neptunium-237:

$$^{241}_{95}\text{Am} \longrightarrow {}^{237}_{93}\text{Np} + {}^4_2\text{He}$$

In all correct nuclear equations, the sum of the mass numbers (superscripts) on the left of the arrow must equal the sum of the mass numbers on the right of the arrow and the sum of the atomic numbers (subscripts) on the left of the arrow must equal the sum of the atomic numbers on the right of the arrow. In the example given:

241 = 237 + 4 (mass numbers)
and 95 = 93 + 2 (atomic numbers)

Summary

When an isotope emits an alpha particle, the atomic number of the isotope decreases by two and the mass number decreases by four.

When an isotope emits a beta particle, the atomic number of the isotope increases by one (because a neutron has been converted into a proton) and the mass number stays the same.

When an isotope emits gamma rays, the atomic number and the mass number of the isotope stay the same.

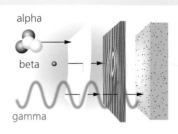

Figure 12.7 The penetrating power of α, β and γ radiation

Questions

3 Copy the diagram below and label the arrows to show which represents α, β and γ radiation. Give a reason for each of your decisions.

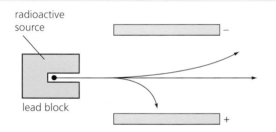

Figure 12.8

4 Construct a table showing the name, symbol, mass, charge and penetrating power of α, β and γ radiation.

5 Which of the changes in the table below takes place when an atom emits:
   a) alpha radiation
   b) beta radiation
   c) gamma radiation?

Atomic number	Mass number
A increased	no change
B no change	increased
C decreased	decreased
D no change	no change

Figure 12.9 Marie Curie (left) and Dorothy Hodgkin (right), two celebrated female scientists

Henri Becquerel discovered radioactivity in 1896. Two of Becquerel's colleagues, Marie and Pierre Curie, carried the work forward and one of them, Marie Curie, was the first person to be awarded two Nobel Prizes for different subjects, one for Physics in 1903 and one for Chemistry in 1911. In June 1903 Marie and Pierre were invited to the Royal Institution in London to give a talk on radioactivity. However because Marie was female, she was not allowed to address the meeting and Pierre alone was allowed to speak. Use the internet to research the lives of some famous women scientists of the twentieth and twenty-first centuries.

Worked examples

1 Name the isotope formed when thorium-232 emits an alpha particle.

$$^{232}_{90}Th \longrightarrow {}^{a}_{b}X + {}^{4}_{2}He$$

In this equation: $232 = a + 4$, so $a = 232 - 4 = 228$; $90 = b + 2$, so $b = 90 - 2 = 88$

If the atomic number is 88, the element is radium, so the isotope is radium-228.

2 Name the isotope formed when lead-212 emits a beta particle.

$$^{212}_{82}Pb \longrightarrow {}^{a}_{b}X + {}^{0}_{-1}e$$

In this equation: $212 = a + 0$ so $a = 212$; $82 = b - 1$, so $b = 83$

If the atomic number is 83, the element X is bismuth, so the isotope is bismuth-212.

3 Name the isotope formed when plutonium-242 captures two alpha particles.

$$^{242}_{94}Pu + 2{}^{4}_{2}He \longrightarrow {}^{a}_{b}X$$

In this equation: $242 + 8 = a$, so $a = 250$; $94 + 4 = b$, so $b = 98$

If the atomic number is 98, the element is californium so the isotope is californium-250.

Questions

6 Find the values of a and b and identify X in each of the following equations:

a) $^{238}_{92}U \longrightarrow {}^{a}_{b}X + {}^{4}_{2}He$

b) $^{32}_{15}P \longrightarrow {}^{a}_{b}X + {}^{0}_{-1}e$

c) $^{220}_{86}Rn \longrightarrow {}^{a}_{b}X + {}^{4}_{2}He$

7 The following sequence of changes occurs as polonium-218 decays.

$$^{218}_{84}Po \xrightarrow{1} {}^{214}_{82}Pb \xrightarrow{2} {}^{214}_{83}Bi$$

Name the type of radiation emitted during changes 1 and 2. Give a reason for each of your decisions.

8 Copy and complete the following nuclear equations:

a) $^{3}_{1}H \longrightarrow {}^{3}_{2}He$

b) $^{224}_{88}Ra \longrightarrow {}^{228}_{90}Th$

c) $^{60}_{27}Co \longrightarrow {}^{60}_{27}Co$

Half-life

Radioactive decay is a random event. There is no way of knowing when any particular atom will decay; it might happen in the next instant or it might only happen after a million years have passed, so the rate of decay of a radioisotope is measured by a period of time known as the **half-life**.

The time it takes for half the atoms of the starting element to change into new isotopes is called the half-life of that element.

After one half-life has passed, 50% of the atoms have changed into new isotopes. For a particular isotope, the half-life is a constant – that is, it takes the same length of time for 1000 atoms to reduce to 500 as it does for 12 atoms to reduce to 6.

Half-life can also be expressed in terms of activity (as measured on a Geiger counter) or the quantity of the original isotope that is left. The half-life of a particular isotope is fixed and will always be the same whether the isotope is in the Sun (very high temperature) or at the bottom of the ocean (low temperature). Figure 12.10 shows the change in activity for an isotope with a half-life of 2 days.

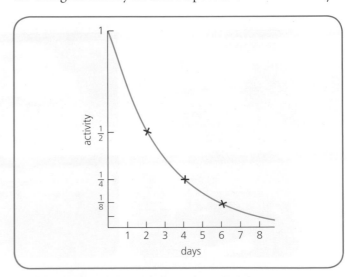

Figure 12.10 Change in activity with time for an isotope of half-life 2 days

Half-lives can vary from seconds to millions of years and are independent of the presence of catalysts or the chemical state of the isotope.

The half-life of the artificial isotope sodium-24 is 15 hours. The particle might be present as an atom,

$^{24}_{11}Na$ or as an ion, $^{24}_{11}Na^+$, but the half-life would still be 15 hours. A bigger mass of sodium-24 will have a greater *intensity* of radiation than a smaller mass, but the two samples will still have the same half-life. Radioisotopes can be used to date materials because the half-life of an individual isotope is constant.

Activities

To examine the random nature of radioactive decay and look at the idea of half-life, place 100 × 2p coins in a plastic beaker, shake the beaker and tip the coins out. Count the number of coins that have landed heads-up, remove them and replace them with 1p coins. (The decayed isotopes have not

vanished – they have decayed into a different isotope, represented by the 1p coins.) Repeat the process, counting the number of heads-up 2p coins after each throw and recording the data. When all the 2p coins have been replaced, plot a graph of the data and discuss how this activity relates to the idea of half-life.

Carbon dating

Cosmic radiation bombards the Earth constantly. Neutrons are part of this cosmic radiation and, in the upper atmosphere, nitrogen atoms capture neutrons to produce the radioisotope carbon-14. A proton is also produced in this reaction.

$$^{14}_{7}N + ^{1}_{0}n \longrightarrow ^{14}_{6}C + ^{1}_{1}p$$

The concentration of carbon-14 atoms in the atmosphere is fairly constant because, as they form, they combine with oxygen and green plants on the planet's surface absorb the CO_2 produced. In living plants, the carbon-14 is slowly but continuously decaying and it is simultaneously being replaced by CO_2 absorbed

during photosynthesis. When a plant dies, it stops taking in carbon dioxide, but the carbon-14 already in the plant keeps on decaying. If we assume that the ratio of carbon-14 to carbon-12 in trees today is the same as it was in the past, then measuring the activity due to carbon-14 in a sample of old wood and comparing it to that of a sample of the same type of new wood enables scientists to estimate the age of the old wood.

Activities

Working in a group, use the internet to find out why the technique of carbon dating cannot be used to find the age of dinosaur bones. Make a poster to show how carbon dating works.

Worked examples

1   Analysis of a piece of cedar found in Jordan showed the carbon-14 content was 6.25% that of living Lebanese cedar. If the half-life of carbon-14 is 5600 years, estimate the age of the sample.

100% ⟶ 50% ⟶ 25% ⟶ 12.5% ⟶ 6.25%
living wood   sample of ancient wood

Each arrow represents one half-life, so four half-lives have passed. The sample is

4 × 5600 = 22 400 years old.

2   Tritium, ^3_1H, is a radioisotope of hydrogen with a half-life of 12.4 years. A bottle of old wine showed an activity of 4 counts min^{-1}, whereas a bottle of new wine from the same vineyard had an activity of 32 counts min^{-1}. Calculate the age of the wine.

32 counts ⟶ 16 counts ⟶ 8 counts ⟶ 4 counts
new wine   old wine

Three half-lives have passed, so the wine is
3 × 12.4 = 37.2 years old.

Questions

9 A luminous watch dial containing $^{147}_{61}$Pm (half-life = 2.5 years) has only 1/8th of its original 'glow'. How old is the watch?

10 The activity of a sample of $^{131}_{53}$I was found to be only 1/16th of the activity when it arrived at the hospital 32 days earlier. What is the half-life of iodine-131?

11 Which of the following processes would alter the half-life of a sample of a radioisotope of magnesium?
A Reacting it with dilute acid
B Cooling it to −10 °C
C Burning it in air
D None of these

Nuclear fission

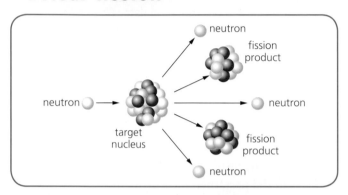

Figure 12.11 Nuclear fission

Nuclear fission (the breaking up of a nucleus) is, at present, the most important way of getting energy from atoms. The main fuel used in nuclear reactors is uranium-235. This is a heavy isotope which can be split into smaller fragments when it is bombarded with neutrons. One pattern of fission is:

$$^{235}_{92}U + ^{1}_{0}n \longrightarrow ^{91}_{36}Kr + ^{142}_{56}Ba + 3^{1}_{0}n$$

The fission reaction releases three more neutrons, which collide with more U-235 nuclei and a chain reaction takes place. When fission occurs, a tiny quantity of mass is converted into energy according to Einstein's famous equation $E = mc^2$, where c = the speed of light = 3×10^8 m s^{-1}. This equation shows that very small quantities of mass are converted into very large quantities of energy. In fact, the fission of about 235 g of uranium generates the same amount of energy as the burning of 60 tonnes of coal, which would release about 220 tonnes of CO_2 into the atmosphere. There are advantages and disadvantages in using nuclear power to generate electricity, but the discussion is often hampered by the association of the peaceful uses of nuclear energy with nuclear weapons.

Activities

Most world leaders agree that we need to lower CO_2 emissions across the planet. Electric cars could be part of the answer to the problem of reducing global warming but charging these vehicles will mean an increasing demand for electricity which will put pressure on present supplies. Sources of renewable energy are being developed but present generation nuclear stations can provide large quantities of 'clean', relatively cheap electrical energy.

Working in a group, use the internet to research arguments for and against nuclear energy. France has over 20 nuclear reactors on-line, providing electricity for domestic and commercial use. The UK buys some electricity from France, but has very few nuclear reactors on the UK grid. Will 'green electricity' fill the gap left by coal-fired power stations? Have a debate on the subject of how we should generate our power in this country.

Nuclear fusion

Uranium is mined from the Earth's crust and is a finite resource, hence there has been a lot of research into energy production by **nuclear fusion**. Fusion reactions create the energy in stars like our Sun, where very light nuclei fuse (combine) together to make heavier nuclei and release even more energy than fission reactions. For example:

$$^{3}_{1}H + ^{2}_{1}H \longrightarrow ^{4}_{2}He + ^{1}_{0}n + energy$$

Fusion has been achieved on this planet, but only for very short time intervals.

Artificial radioisotopes

Radioisotopes (or radionuclides) are manufactured for use in many industries. These can be made by bombarding stable isotopes with neutrons in a nuclear reactor. Neutrons have no charge, so they are not repelled by the positive nucleus.

For example:

$$_{13}^{27}\text{Al} + _{0}^{1}\text{n} \longrightarrow _{11}^{24}\text{Na} + _{2}^{4}\text{He}$$

Uses of radioisotopes

Medicine

1 Cobalt-60 is an artificially produced isotope that carries excess energy. It emits γ-rays to form a lower energy version of the same isotope, and is one of several isotopes used in radiotherapy to destroy cancer cells.
2 Many isotopes are used to trace the movement of materials in our bodies. Salt solution containing a small amount of sodium-24 can be injected into the bloodstream to find out whether blood circulation is normal.
3 Radioisotopes of iodine can be used to test how well the thyroid gland is working, and iodine-131 can be used in radiation therapy for the treatment of thyroid cancer.

Industry

1 Domestic smoke alarms use the α-emitter, americium-241. This isotope is used because the alarms are safe (α-particles have a low penetrating power so are absorbed by the plastic covering) and the isotope has a long half-life (>400 years) so it does not need to be replaced.
2 Beta emitters are widely used to monitor the thickness of materials like paper, plastic and thin metal sheets. Gamma emitters are used to examine castings and welds for imperfections.
3 In many countries food is irradiated, in other words exposed to electron beams, X-rays or gamma rays. The process can be used to kill bacteria that cause food poisoning, such as salmonella and *E. coli*. It can also delay fruit ripening and help stop vegetables such as potatoes and onions from sprouting. Irradiation has little effect on the look and texture of food and though irradiated food has been exposed to radioactivity, it does not become radioactive itself. In the same way that clothes on a washing line exposed to the Sun's rays do not hold the light energy and start to shine, the treated food carries no trace of the high-energy beams it was exposed to.
4 Radioisotopes are used to detect leaks in pipelines and to trace underground water. On big farms, they are often used to check for proper distribution of insecticides.

Checklist for Revision

- I can identify the various radiation processes, alpha, beta and gamma decay.
- I can state the mass, charge and penetrating power of alpha, beta and gamma radiation.
- I can write nuclear equations.
- I can give some uses of radioisotopes.
- I understand the concept of half-life and can use it to solve problems.

End-of-chapter questions

1* Carbon dating uses carbon-14 isotopes and can be used to estimate the age of charcoal found in archaeological sites.
   a) Carbon-14 decays by beta emission. Write a balanced nuclear equation for this decay.
   b) Why can carbon dating *not* be used to estimate the age of fossil fuels?

2 Sulfur-35 is a radioisotope that decays by beta emission.
   a) Write the nuclear equation for the decay of sulfur-35.
   b) The half-life of sulfur-35 is 87 days. How long would it take for the mass of sulfur-35 in an 8 g sample to fall to 2 g?

3 Find the values of a and b and identify X in each of the following equations:
   a) $^{234}_{90}\text{Th} \longrightarrow {}^{a}_{b}\text{X} + {}^{0}_{-1}\text{e}$
   b) $^{210}_{84}\text{Po} \longrightarrow {}^{a}_{b}\text{X} + {}^{4}_{2}\text{He}$
   c) $^{214}_{82}\text{Pb} \longrightarrow {}^{a}_{b}\text{X} + {}^{0}_{-1}\text{e}$

4 Radioisotopes are used in medicine and in industry.
   a) Name a radioisotope used to treat cancer by radiotherapy.
   b) Give two reasons why americium-241 is suitable for use in domestic smoke detectors.
   c) Explain why a γ-emitting radioisotope could not be used to check the thickness of paper as it comes off the rollers in a paper mill.

5 Explain how carbon-14 isotopes are formed and how these isotopes can be used to date samples of wood found in archaeological sites.

6 Everyone on this planet is exposed to background radiation from various sources.
   a) Name one source of background radiation.
   b) Where were all the naturally occurring elements on this planet formed? (You may need to carry out some internet research to answer this question.)

13 Chemical analysis

Common chemical apparatus

The National 5 chemistry course involves lots of practical work and by now, you should be familiar with the use of many pieces of common chemical apparatus, such as beakers, test tubes and boiling tubes, balances, measuring cylinders, conical flasks, droppers, filter funnels and papers, evaporating basins, pipettes and automatic pipette fillers, burettes and thermometers among others. In this chapter, we will look at the use of this apparatus in some situations which have not been covered already in the book. It is also important to remember that experimental chemistry is not performed solely out of the interest of the experimenter – there are benefits in the 'real world' arising from the knowledge gained by carrying out practical work. For example, you will be aware from hearing or reading the news that people are concerned about the damage that our twenty-first century lifestyle is causing to 'the environment'. What does that mean and how can chemistry help?

Consumption and the environment

In the 66 years between Orville Wright piloting the first successful powered flight in 1903 to Neil Armstrong stepping onto the Moon in 1969 (Figure 13.1), scientists and engineers have transformed and improved our lives beyond belief. Science and scientists have given us cheap electricity and gas to make our lives warm and comfortable and, more importantly, cheap energy has encouraged the discovery and mass production of new materials. Factories manufacture (and package) consumer goods, and if the factories are to continue manufacturing, the consumers have to buy more goods. As soon as we started throwing things away, rather than mending them, we created environmental problems and those problems appear, at present, to be increasing and multiplying. Scientists create solutions

Figure 13.1 Orville Wright's first successful flight; Neil Armstrong stepping out onto the Moon

to problems, but it is up to society whether or not those solutions are accepted. Bio-plastics are being made from corn, rice, potatoes and sugar cane, but these polymers cost a little more and are not always as convenient as traditional oil-based polymers. We consumers like our goods to be cheap and convenient, so our environment continues to be littered with oil-based plastics that can take centuries to decompose.

Water and pollution

Figure 13.2 Loch Katrine in the district of Stirling

Figure 13.3 Water pollution can have devastating effects on fish

The increasing emphasis on recycling is helping to reduce the mountains of solid material we throw away each year, but water quality is an on-going problem. Scottish Water delivers our clean drinking water (Figure 13.2) and maintains the sewers, drains and water treatment plants. Chemists are employed to monitor and maintain water quality at every step of this efficient process, but industry also needs water.

Distilleries, food processing plants, farms, paper mills, glass factories and just about every other manufacturing industry all use a great deal of water; they then have to clean it before they dispose of it. The Scottish Environmental Protection Agency (SEPA) is the organisation that regulates emissions to our environment. Discharges of water from industrial sites and agriculture are controlled by regulations. Large sites are controlled by licences which limit (among other things) the temperature, pH, conductivity, total suspended solids and the concentrations of nitrate, sodium, potassium, calcium, magnesium, iron, manganese, phosphate and sulfate ions in this waste water. If a manufacturer breaches this licence, they can be prosecuted through the courts. This means that both SEPA and the manufacturing companies employ chemists to continually check the quality of the water that is being discharged and to 'trouble-shoot' when a problem is found.

Causing and controlling pollution

We often hear people say they don't like 'things with chemicals in them'. This is an odd statement because the

substance most essential to life, water, is a chemical and human beings are a mixture of thousands of chemicals, each doing a specific job in our metabolic system. What they mean is that none of us want to find an excess of a particular chemical in a place where it would not occur naturally. For example, how can water become polluted? Here are just two ways:

- Artificial fertilisers can be washed off fields during heavy rain and cause the rapid growth of algae in ponds and lochs. Algae are short-lived. When they die and decay, the decaying process removes oxygen from the water. Fish 'breathe' dissolved oxygen, so the presence of algae can make it difficult for fish to survive.
- A variety of metal compounds can leach into the water from old flooded mine workings.

Each industry has its own type of waste material to add to discharged water. Chemists have developed ways to remove all the unwanted compounds from water, so as long as each industry behaves responsibly, pollution can be avoided.

Techniques for monitoring the environment

Apparatus can be taken out of the lab and used to analyse, for example, samples of water or soil taken on a site. Usually, however, samples are sent to professional labs for detailed and accurate analysis. Although much of this analysis is automated, the chemical techniques and analytical methods used are often pretty similar to the ones used in school labs.

General practical techniques

Earlier in this book, you learned how to:

- use a balance (Chapter 1, page 4)
- collect a gas by downward displacement of water (Chapter 1, page 5)
- test the electrical conductivity of substances (Chapter 3, page 21)
- prepare soluble salts by titration or neutralisation of an acid by an insoluble base (Chapter 5, pages 42 and 43)
- heat substances using a Bunsen burner (Chapter 5, page 43)
- separate an insoluble solid from a liquid by filtration (Chapter 5, page 43)
- measure the energy released when a fuel burns (Chapter 8, pages 75–77)
- perform an electrolysis experiment (Chapter 9, pages 89 and 91)
- set up electrochemical cells in various ways (Chapter 9, pages 92–94).

In this chapter, you will learn how to:

- collect gases that are soluble in water
- prepare insoluble salts
- heat flammable substances that cannot be safely heated using a Bunsen burner.

Collecting gases that are insoluble in water

Gases, such as hydrogen, oxygen, carbon dioxide and other relatively insoluble gases, can be collected by downward displacement of water in an experiment using the method shown in Figure 13.4.

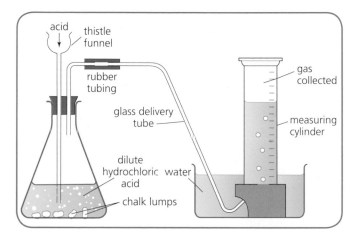

Figure 13.4 Collecting a gas by downward displacement of water

A gas syringe can also be used to do this, as shown in Figure 13.5.

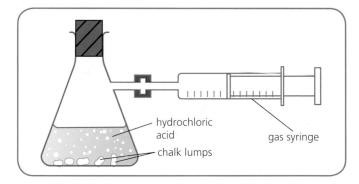

Figure 13.5 Collecting a gas in a gas syringe

If we want to measure the volume of gas produced, then the gas has to be collected in a vessel that is marked off in volume graduations, such as an inverted measuring cylinder filled with water or a graduated syringe. Note that when using the apparatus shown in Figure 13.4, it is important to make sure that the end of the funnel used to add a liquid reactant to the reaction vessel is beneath the final surface of the liquid so that there is only a single exit route for the gas produced. If volume does not need to be measured, then the gas can be collected in an upturned test tube or boiling tube filled with water.

If a gas is soluble in water (for example ammonia or sulfur dioxide), the first method cannot be used – the gas would dissolve in the water rather than push it down and out of the receiving vessel. The syringe method can be used, but there is an alternative method which depends on the density of the gas.

Collection of a gas by displacement of air

As a rough guide, if a gas has a gram formula mass of less than 30 g, it is less dense than air and can be collected by downward displacement of air. If the GFM is greater than 30 g, upward displacement of air can be used.

Collecting a gas by downward displacement of air

Ammonia (NH_3) has a GFM of 17 g and so it is less dense (lighter) than air; ammonia will push air down from an inverted collection vessel (see Figure 13.6, downward displacement of air).

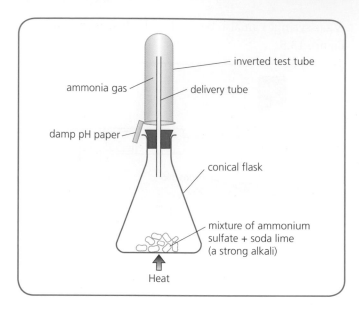

Figure 13.6 Collecting ammonia by downward displacement of air

Damp pH paper will turn blue when the collecting tube is full.

Collecting a gas by upward displacement of air

Sulfur dioxide (SO_2) has a GFM of 64 g and so it is denser than air; sulfur dioxide will push air up from a collection vessel (see Figure 13.7, upward displacement of air).

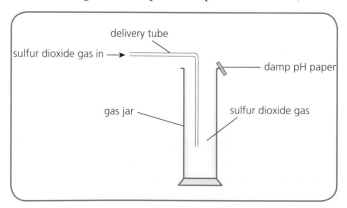

Figure 13.7 Collecting sulfur dioxide by upward displacement of air

Damp pH paper will turn pink when the gas jar is full.

Preparing insoluble salts by precipitation

In Chapter 5, you learned that salts are formed following the replacement of the hydrogen ions of an acid by metal ions or the ammonium ion. Soluble salts can be prepared by neutralising an acid with a base. The SQA National 5 Data Booklet will tell you whether or not a

salt is soluble. For insoluble salts, it is not necessary to start with an acid. They can be prepared by **precipitation**. Precipitation is the reaction of two solutions to form an insoluble substance called a precipitate.

If we look at the key to the solubility table in the SQA National 5 Data Booklet, we can see that insolubility is a matter of degree. 'Insoluble' means 'a solubility of less than 1 g per litre' and that is a low enough solubility for us to make the general statement 'the compound is insoluble in water'. These compounds are insoluble because the attraction of the positive and negative ions inside the ionic lattice (see Chapter 3) is very strong. This fact makes it easy to prepare insoluble salts. All we need to do is identify the positive and negative ions in the required salt then bring these ions together in a solution. As the ions collide they will tend to 'stick' together and the strong lattice structure will start to form. The particles of solid precipitate will rapidly get bigger and turn the solution cloudy. Precipitation is a very fast reaction. To obtain a sample of the pure, dry insoluble salt, the precipitate can be filtered off, washed (to remove traces of the soluble salt that remain in the solution) and dried.

This technique can be used to prepare a sample of the insoluble salt, silver(I) iodide, Ag^+I^-, as described in Figure 13.8. For practical purposes in the laboratory, we generally use soluble compounds marked in the Data Booklet as '*vs*' (very soluble) when we are preparing insoluble salts by precipitation.

First, we use the SQA Data Booklet to identify a very soluble salt that contains silver(I) ions, for example silver(I) nitrate, and a very soluble salt which contains iodide ions, such as sodium iodide. If we make up solutions of both compounds and then mix them, the following reaction occurs:

silver(I) nitrate(aq) + sodium iodide(aq) $\longrightarrow$
silver(I) iodide(s) + sodium nitrate(aq)

$Ag^+(aq) + NO_3^-(aq) + Na^+(aq) + I^-(aq) \longrightarrow$
$Ag^+I^-(s) + Na^+(aq) + NO_3^-(aq)$

If we take out the spectator ions, the reaction can be written as follows:

$$Ag^+(aq) + I^-(aq) \longrightarrow Ag^+I^-(s)$$

Details of the practical technique are shown in Figure 13.8.

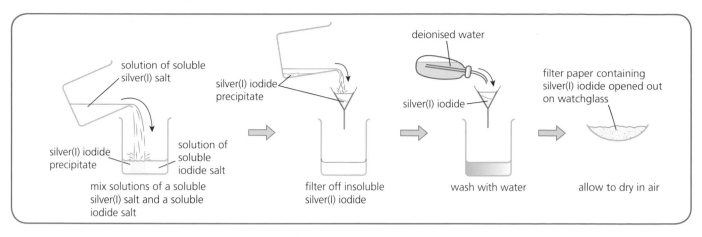

Figure 13.8 Preparing the insoluble salt, silver(I) iodide, by precipitation

Questions

1 Name the insoluble product formed when aqueous solutions of the following are mixed:
   a) nickel(II) chloride and potassium carbonate
   b) barium chloride and sulfuric acid.
2 Lead(II) carbonate is a white, insoluble salt.
   a) i) Name two solutions that could be mixed to obtain lead(II) carbonate.
   ii) Write a balanced chemical equation, including state symbols, for this reaction.
   iii) Rewrite the equation leaving out the spectator ions (include state symbols).
   b) What name is given to this type of reaction?
3 Draw a labelled, sectional diagram to show the apparatus that can be used to separate an insoluble solid from a solution.

Heating without a Bunsen burner

The final step in preparing an insoluble salt is usually to dry the salt on filter paper in an oven after filtration. If the salt is soluble, the final step is to evaporate water from the mixture of salt and water to produce a dry sample of the salt. This is usually done with a Bunsen burner. As you learned in Chapter 7, many carbon compounds, such as alcohols, are flammable. Yet sometimes when carrying out chemical reactions with such compounds, heat is required. When a Bunsen burner cannot be used, chemists have alternative methods.

When using a hotplate or water bath, the temperature can be set using the thermostat control so that it is below the temperature at which any flammable chemicals will evaporate and possibly catch fire. A quick

Figure 13.9 Heating without a Bunsen burner

alternative is to put hot water from a kettle in a beaker and use a thermometer to check its temperature. Again, provided the temperature is below the boiling point of any flammable reactants, a boiling tube containing the reaction mixture can safely be placed in the beaker. You may have used this method to remove chlorophyll from green leaves using ethanol earlier in your school career.

Analytical methods

Tests for some common gases

Oxygen: The test for oxygen is that it relights a glowing splint. Air does not give this result because it does not contain enough oxygen.

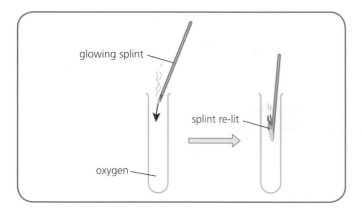

Figure 13.10 Oxygen re-lights a glowing splint

Hydrogen: When hydrogen is mixed with air and a lighted splint or taper is applied, the hydrogen burns with a pop. This reaction is used as the test for hydrogen.

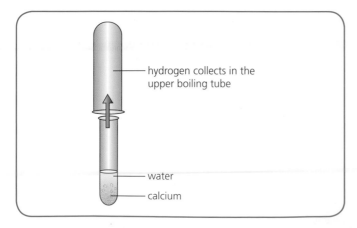

Figure 13.11 Hydrogen can be collected by downward displacement of air

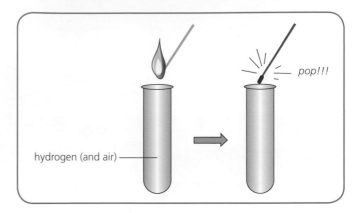

Figure 13.12 Hydrogen burns with a pop

Carbon dioxide: The test for carbon dioxide is that if the gas is bubbled through limewater, the limewater turns cloudy. Limewater is a saturated solution of calcium hydroxide (a base) so carbon dioxide (an acidic gas) reacts with the base to form the insoluble salt, calcium carbonate.

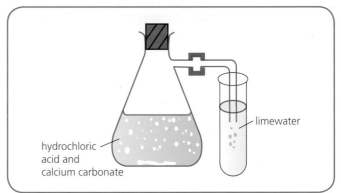

Figure 13.13 Carbon dioxide turns limewater cloudy

Flame tests

Some metal compounds will produce a distinctive colour when they are held at the edge of a blue Bunsen flame (Figure 13.14). The colour is due to the presence of the metal ions in the compound. Some of these colours are shown in Table 13.1.

The Royal Society of Chemistry (RSC) has worked with the Nuffield Foundation to produce resources for practical school chemistry. Search the RSC website for a demonstration of the technique of flame testing.

Element	Ion	Colour
barium	Ba^{2+}	green
calcium	Ca^{2+}	orange-red
copper	Cu^{2+}	blue-green
lithium	Li^+	red
potassium	K^+	lilac
sodium	Na^+	yellow
strontium	Sr^{2+}	red

Table 13.1 Some common flame test results

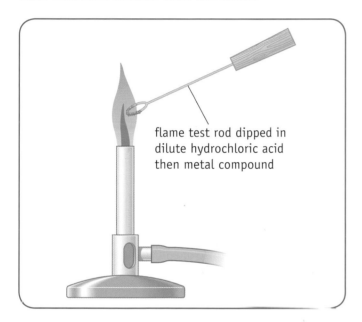

flame test rod dipped in dilute hydrochloric acid then metal compound

Figure 13.14 Flame test

Chemical tests for some positive ions in aqueous solution

Most metal hydroxides are insoluble in water and are formed as precipitates when sodium hydroxide solution is added to a solution containing the metal ion. Several metals can be identified by the colour of their insoluble hydroxides, as shown in Table 13.2.

Metal ion in solution	Precipitate colour on adding a few drops of sodium hydroxide solution
aluminium	white, but precipitate re-dissolves in excess sodium hydroxide solution
iron(III)	brown
copper(II)	blue
iron(II)	green
magnesium	white

Table 13.2 Identifying metals by the colour of their insoluble hydroxides

(The RSC microscale experiments can be used to minimise the quantities of reactants used for these tests. Details of the microscale procedures can be found in the RSC book *Inspirational Chemistry – Resources for Modern Curricula*, 2006, sections 7.1 to 7.3.)

Chemical tests for some negative ions in aqueous solution

Some common tests for negative ions are shown in Table 13.3.

Ion	Test	Result
chloride, Cl^-	Add a few drops of dilute nitric acid then a few drops of silver(I) nitrate solution.	White precipitate of silver(I) chloride is formed. (Check: this precipitate should be soluble in dilute ammonia solution.)
bromide, Br^-		Cream/off-white precipitate of silver(I) bromide is formed.
iodide, I^-		Yellow precipitate of silver(I) iodide is formed.
carbonate, CO_3^{2-}	Add dilute hydrochloric acid to a solid (or to a solution) containing carbonate ions.	Effervescence (bubbling/fizzing); colourless gas given off – the gas is carbon dioxide, it turns limewater cloudy.
sulfate, SO_4^{2-}	Add a few drops of hydrochloric acid then a few drops of barium chloride solution.	White precipitate of barium sulfate formed – this precipitate is made of very small, fine particles.
nitrate, NO_3^-	Add a few drops of sodium hydroxide solution and a little aluminium powder. Warm the solution in a Bunsen flame and test any gas given off using damp pH paper.	A gas is given off which turns the damp pH paper blue. This shows that the gas is ammonia.

Table 13.3 Some common tests for negative ions

Preparing and using a standard solution

In Chapter 5, you learned that a standard solution is a solution of accurate and known concentration. Imagine you were asked to find out the concentration of some hydrochloric acid solution. What you would need to do is prepare a standard solution of a base and then use it in a series of titrations. The results of the titrations would enable you to calculate the concentration of the acid solution.

A suitable base for this purpose would be sodium carbonate, Na_2CO_3. It is a very soluble solid which can be obtained in a very pure state, meaning that when a solution is made from it, it will contain no impurities. To make a standard solution, you need to calculate and then weigh accurately the required mass of solute (in this case, sodium carbonate) on an electronic balance. If you wanted to make $250\,cm^3$ of a $0.1\,mol\,l^{-1}$ solution, you would have to first calculate the number of moles of solid required:

Number of moles (n) = concentration (C) × volume in litres (V)

$$= 0.1 \times 0.25$$
$$= 0.025 \text{ moles}$$

Then, you would have to calculate the mass of solid required:

Mass in grams = number of moles × mass of 1 mole of Na_2CO_3

$$= 0.025 \times 106\,g \quad \text{[obtained from } 23 + 23 + 12 + (3 \times 16)]$$
$$= 2.65\,g$$

This mass of pure sodium carbonate would then be weighed accurately using an electronic balance. The next step is to dissolve the solid in **deionised water** to produce a final solution with an extremely accurate volume. This is achieved through using a special piece of glassware known as a **standard flask**. Standard flasks come in many sizes: $50\,cm^3$, $100\,cm^3$, $250\,cm^3$, $500\,cm^3$, 1 litre and even 2 litre capacity (see Figure 13.15). However, they are all the same shape.

Each flask has only a single volume marking (the calibration mark) on its neck. They can measure that volume, and only that volume, with great accuracy. To use a standard flask, the following procedure should be followed:

Figure 13.15 Standard flasks

- Weigh out accurately the required mass of solute (in this case, $2.65\,g$ of pure sodium carbonate) into a clean beaker.
- Add sufficient deionised water and stir the mixture until all of the solid has dissolved.
- Transfer this solution to a standard flask (in this case, a $250\,cm^3$ flask) using a clean filter funnel.
- Rinse the beaker several (at least three) times with more deionised water and transfer the rinsings to the flask using the funnel.
- When the total volume of solution has reached the bottom of the neck of the flask, top up the solution using deionised water from a plastic wash bottle.
- When the bottom of the solution meniscus is on the calibration mark, the volume is exactly $250\,cm^3$.
- At that point, stopper the flask and invert it several times to ensure that the concentration of the solution is uniform throughout. If you do not do this, the solution will be more concentrated near the bottom and less concentrated near the top.

You have now made a standard solution. The only solute in the solution is sodium carbonate because you used a very pure sample of it and used deionised, rather than tap, water. The concentration of the solution will be $0.1\,mol\,l^{-1}$. You would now be able to perform several titrations on the acid solution of unknown concentration until you had two concordant results. From there, using the equation for the reaction:

$$Na_2CO_3(aq) + 2HCl(aq) \longrightarrow 2NaCl(aq) + H_2O(l) + CO_2(g)$$

you can calculate the concentration of the acid solution (see Chapter 5, pages 43 and 44).

1   Look at the tables of tests (and results) given in this chapter and choose a combination of positive and negative ions to make a compound. On a piece of scrap paper, write down the correct formula for your compound and the results of two tests that will identify the positive and negative ions present in your compound. Ask your teacher/lecturer to check the formula you have written and keep the paper, so you can check the answers.

Copy the 'test results' neatly onto a second piece of paper, fold the paper up and put it into a beaker with the test results of the rest of the class. Each member of the class has then got 30 seconds to take a piece of paper from the beaker, identify the compound from the test results and give the correct formula, before the beaker passes on.

2   If time allows, your teacher/lecturer may give you an unknown sample, which might contain a single compound or it might contain a mixture of two compounds. Your task is to plan, then carry out, a series of experiments to identify the ions present in the sample and suggest a name for the compound(s) present.

First, write down an aim for your experiment; this is what you are trying to do (in this case, you are trying to identify an unknown compound). Then write a plan to show the order in which you are going to perform each test and decide how you are going to record the observations. Think about how you are going to ensure your experiments are safe. Ask your teacher/lecturer to check your plan.

When you have finished the practical work, write down what you have found out (this usually goes under the heading 'Conclusion'), then think about how you could have made the process more efficient. If you were going to do the experiment again, would you have done the tests in the same order? Did you make any mistakes? Do you think you correctly identified your compound? If not, why not?

Reporting experimental work

When experimental work is done in any science, a report is written so that the work can be repeated and checked by other scientists working in the same field. Skills and knowledge that are required for report writing include:

- drawing labelled, sectional diagrams for common chemical apparatus
- presenting data in tabular form with appropriate headings and units of measurement
- presenting data as a bar, line or scatter graph with suitable scale(s) and labels
- using a line of best fit (straight or curved) to represent the trend observed in experimental data
- calculating average (mean) values from data
- suggesting and justifying an improvement to the experimental method given a description of an experimental procedure and/or experimental results.

Your teacher/lecturer will provide opportunities for you to learn and practise these skills as you progress through the National 5 course. Several of the skills are used during the report writing stage of the following experiment, one which you may have done in your earlier school years.

An experiment to investigate the effect of particle size on rate of reaction

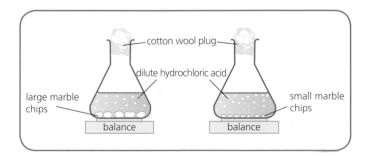

Figure 13.16 Labelled, sectional diagram of apparatus needed to investigate the effect of particle size on rate of reaction

The mass of each flask is measured every minute and the raw data collected is shown in Table 13.4.

Time (minutes)	Mass of apparatus (g)	
	Small marble chips	Large marble chips
0	198.1	197.9
1	196.6	197.0
2	195.4	196.1
3	194.6	195.4
4	194.0	194.7
5	193.5	194.1
6	193.3	193.6
7	193.1	193.3
8	193.1	193.0
9	193.1	192.9
10	193.1	192.9

Table 13.4 Raw data

Time (minutes)	Mass of carbon dioxide produced (g)	
	Small marble chips	Large marble chips
0	0.0	0.0
1	1.5	0.9
2	2.7	1.8
3	3.5	2.5
4	4.1	3.2
5	4.6	3.8
6	4.8	4.3
7	5.0	4.6
8	5.0	4.9
9	5.0	5.0
10	5.0	5.0

Table 13.5 Processed data

Both variables are continuous so it is appropriate to present the data in Table 13.5 as a pair of line graphs and to draw (curved) lines of best fit to represent the trends observed.

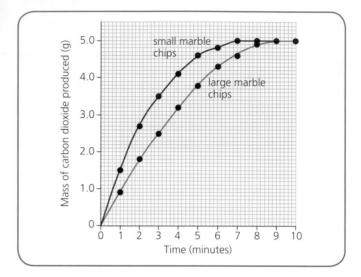

Figure 13.17 Line graphs showing trends observed in the processed data

Questions

4　Consider the apparatus shown in Figure 13.16.
   a) Why do the flasks get lighter as time progresses?
   b) The reaction between acid and marble chips starts as soon as the reactants are mixed. Suggest how another piece of apparatus could be used to enable the mass (of apparatus + reactants) to be accurately measured at time zero.
5 a) Explain how the processed data in Table 13.5 was obtained from the raw data in Table 13.4.
   b) What is meant by the term 'continuous variable'?
   c) Why do both lines finally become horizontal?
   d) Calculate the average rate of reaction over the first three minutes using:
      i) the small marble chips
      ii) the large marble chips.

Checklist for Revision

- I am familiar with the use of some common types of chemical apparatus.
- I am familiar with some general practical techniques used in chemistry.
- I know simple tests that can be used to identify oxygen, hydrogen and carbon dioxide gases.
- I know some analytical methods that can be used to monitor the environment.
- I am familiar with the technique of acid/base titration and can perform calculations based on the results of titrations.
- I can report on experimental work in an appropriate manner.

End-of-chapter questions

1 Use the data presented in Table 13.6 to identify compounds A, B, C and D.

Compound	Colour of aqueous solution	Addition of a few drops of acid then of silver(I) nitrate solution	Addition of a few drops of acid then of barium chloride solution	Additional information
A	blue	no reaction	white precipitate formed	blue-green flame test
B	colourless	white precipitate formed	no reaction	lilac flame test
C	colourless	no reaction	white precipitate formed	pH of solution = 1
D	colourless	no reaction	white precipitate which is soluble in dilute ammonia solution	compound D is very soluble in water

Table 13.6

2 A sample of a blue compound, containing copper(II) ions and sulfate ions, was dissolved in water. The solution was divided into three portions and each portion was tested separately. The following results were recorded:

A Sodium hydroxide solution was added and a blue precipitate was formed.

B A few drops of hydrochloric acid were added followed by a few drops of barium chloride solution.

C The compound gave a blue-green colour in a flame test.

a) Write the formula for the precipitate formed in test A.

b) Describe what would be seen in test B.

c) Describe briefly how you would carry out a flame test.

3 Some school students collected a sample of water from their local river and tested it with universal indicator paper; the pH of the water was approximately 10. The students took the sample back to the laboratory and titrated it against a dilute acid because they wanted to find the concentration of hydroxide ions in the water.

a) Search the internet to find the name of a piece of laboratory equipment that could be used to give a more accurate pH reading.

b) i) Name the two pieces of apparatus that are used to measure accurate volumes during a titration.

ii) Use a pencil and ruler to draw a diagram of the apparatus you would assemble in order to titrate a sample of river water. Label each piece of apparatus and each solution.

iii) How would you know when the river water had been neutralised?

c) The results obtained during the titration are shown in Table 13.7. Use the concordant titres to calculate the average titre for this experiment.

	Rough titre	1st titre	2nd titre
Initial burette reading (cm^3)	0.0	15.3	29.8
Final burette reading (cm^3)	15.3	29.8	44.4
Volume used (cm^3)	15.3	14.5	14.6

Table 13.7

d) This experiment is unlikely to give an accurate value for the concentration of hydroxide ions in the sample. Can you identify a possible source of error in the students' reasoning? Give an explanation for your answer.

Appendix

Open-ended questions

Open-ended questions are designed to test your depth of understanding of chemistry. There will be two of these questions in Section 2 of the National 5 examination, each of which is worth 3 marks. In preparing for the course assessment, the following advice may be useful:

- In National 5 Chemistry examinations, open-ended questions will *always* be identified by the wording '**Using your knowledge of chemistry**, …'.
- Read the question carefully. Pay attention to diagrams, structural formulae, equations etc. that have been included in the question to help you answer it.
- Reflect on the information provided in the question. Make sure that you answer exactly what the question is asking.
- You may choose to present your answer as a paragraph or a set of bullet points, and you may also include a diagram.
- Show your understanding of chemistry by explaining (rather than simply stating) chemical information, perhaps by drawing structural formulae, identifying functional groups, writing chemical equations or working out formulae.
- There will not be a single 'correct' answer to an open-ended question. Markers will reward your understanding of chemistry.

Marking

In each of the following marking descriptions, a key phrase is 'understanding of the chemistry involved'. It is not the case that, for example, giving two correct pieces of information will result in the award of 2 marks.

1 mark: The student has:

- demonstrated a limited understanding of the chemistry involved;
- made some statement(s) which is/are relevant to the situation, showing that at least a little of the chemistry within the problem is understood.

2 marks: The student has:

- demonstrated a reasonable understanding of the chemistry involved;
- made some statement(s) which is/are relevant to the situation, showing that the problem is understood.

3 marks: The student has:

- demonstrated a good understanding of the chemistry involved;
- shown a good comprehension of the chemistry of the situation and provided a logically correct answer to the question posed;
- possibly included a statement of the principles involved, a relationship or an equation, and the application of these to respond to the problem.

Examples and model solutions

In each of the following three examples, what follows are simply suggestions. Any correct, relevant chemistry can be awarded marks.

Example 1

The Periodic Table groups together elements with similar properties. In most Periodic Tables, hydrogen is placed at the top of Group 1, but in some it is placed at the top of Group 7.

Using your knowledge of chemistry, comment on the reasons for hydrogen being placed either above Group 1 or Group 7.

Planning your answer

To answer a question like this, focus on chemistry points that you can make that relate to the National 5 course. In Chapters 2 and 3 of this book, you learned much that can be used to answer this question. You learned about the Periodic Table and how it is laid out, what Groups are, the names of some Groups, the common properties of elements in a Group, atoms and molecules, covalent bonding in elements and compounds, intramolecular and intermolecular bonds, ions and ionic bonding.

You do not need to decide in which of the two Groups hydrogen truly belongs, rather you should discuss points in favour of the inclusion of hydrogen in either Group. Think about what belonging to Group 1 or Group 7 tells you about each of the elements in those Groups. What properties do these elements share with hydrogen?

Solution

Group 1

Group 1 elements are reactive because their atoms have one electron in their outer energy level. Hydrogen is also reactive because its atoms only have one electron, and so it could be placed in Group 1.

Atoms of Group 1 elements form ions with a 1+ charge as a result of losing their single outer electron. For example, $Na \longrightarrow Na^+ + e^-$.

You may also include diagrams showing the electron arrangements to supplement your answer as an example of the similarities:

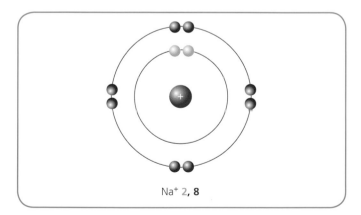

Na$^+$ **2, 8**

Figure 1 How the electrons are arranged in a sodium ion

Hydrogen atoms also form ions with a 1+ charge when they lose their only electron: $H \longrightarrow H^+ + e^-$, and so it could be placed in Group 1.

Group 7

Atoms of Group 7 elements share electrons to form diatomic molecules (e.g. F_2, Cl_2) with intramolecular covalent bonds.

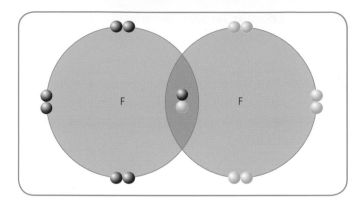

Figure 2 Bonding in a diatomic fluorine molecule

Hydrogen also exists as diatomic molecules (H_2) with intramolecular covalent bonds, and so it could be placed in Group 7.

Atoms of Group 7 elements complete their outer energy level of electrons by gaining one electron. This is also true of hydrogen atoms.

Example 2

The mass spectrometer is a machine used to find out information about the isotopes that elements have. Part of the experimental process involves atoms of the analysed element being converted to positive ions.

Using your knowledge of chemistry, comment on whether the positive ions produced would be larger, smaller or the same size as the atoms from which they were made.

Planning your answer

To answer a question like this, focus on chemistry points that you can make that relate to the National 5 course. In Chapter 2 of this book, you learned about the internal structure of atoms (atomic structure, atomic number, mass number, protons, neutrons, electrons, electron arrangement) and that not all of the atoms of any element are exactly the same due to the existence of isotopes. You also learned that atoms are neutral but that they can be converted into charged particles called ions by losing or gaining electrons.

Since the nucleus of an atom is extremely small, the size of an atom or ion will largely be determined by the number of electron energy levels it has (the greater the number of energy levels, the larger the atom or ion is likely to be) and the extent to which these are pulled in towards the positive nucleus (the greater the positive charge on the nucleus, the smaller the atom or ion is likely to be).

Listed below are some of the concepts that may be included and explained in your answer. This list is not exhaustive and you do not have to include all of these to gain the full 3 marks.

Ensure that what you have included has been explained fully and clearly. Include all diagrams, equations etc. that may help with your explanation.

- Definition of isotopes
- Atomic number and mass number
- Definition of ions
- Ion–electron equations

Solution

Isotopes are atoms with the same atomic number but different mass numbers. This means that they have the same number of protons and electrons but a different number of neutrons.

Atoms are neutral because they have equal numbers of positive protons and negative electrons. When an atom gains or loses electrons, a charged particle called an ion is formed. If an atom gains electrons a negative ion is formed and if an atom loses electrons a positive ion is formed.

Positive ions formed in the mass spectrometer would be smaller than the atoms from which they were formed. If all of an atom's outer electrons are lost, there will be one less occupied energy level in the ion. For example, when a sodium atom loses an electron:

$$Na \longrightarrow Na^+ + e^-$$

Electron arrangement: $2,8,1 \longrightarrow 2,8$

the ion will be smaller because it only has two occupied energy levels of electrons instead of three, as in the atom.

Additionally, the ion still has 11 protons (the atomic number has not changed, it is still 11) in the nucleus but only 10 electrons orbiting it whereas the atom has 11 protons and 11 electrons. This results in a greater force of attraction between the nucleus and the electrons in the ion compared to the atom, making it smaller.

Non-metal elements usually form negative ions by gaining electrons but in the mass spectrometer, they are caused to form positive ions. Again, these ions will be smaller than the atoms due to the protons in the nucleus being able to pull more strongly on a smaller number of electrons.

Example 3

Many different hydrocarbon compounds exist. A student was given three bottles labelled: A (C_6H_{12}), B (C_6H_{12}) and C. The chemical formula for bottle C was not stated on the label. The student was told that one bottle contained hexane, another contained hex-1-ene and the third bottle contained cyclohexane.

Using your knowledge of chemistry, describe how the student could correctly identify the contents of each bottle.

Planning your answer

To answer a question like this, focus on chemistry points you can make that relate to the National 5 course. In Chapter 6 of this book, you learned what hydrocarbons are and how they are grouped in homologous series according to certain criteria. You learned about three of these series – the alkanes, the cycloalkanes and the alkenes. You learned how to represent hydrocarbons using molecular formulae, short and full structural formulae, and that many hydrocarbons have isomers. You also learned about some of the reactions that hydrocarbons can undertake.

A good way to start an answer to a question like this is to demonstrate what you know about each of the three compounds.

Solution

Hexane is an example of an alkane and has the formula C_6H_{14}. Its structure is:

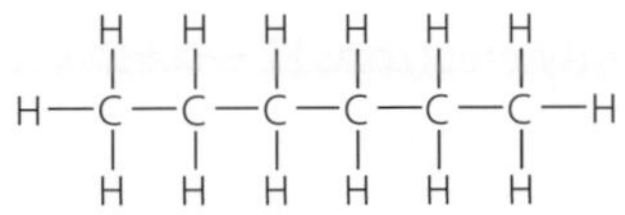

Figure 3 Hexane

Hex-1-ene is an example of an alkene and has the formula C_6H_{12}. Its structure is:

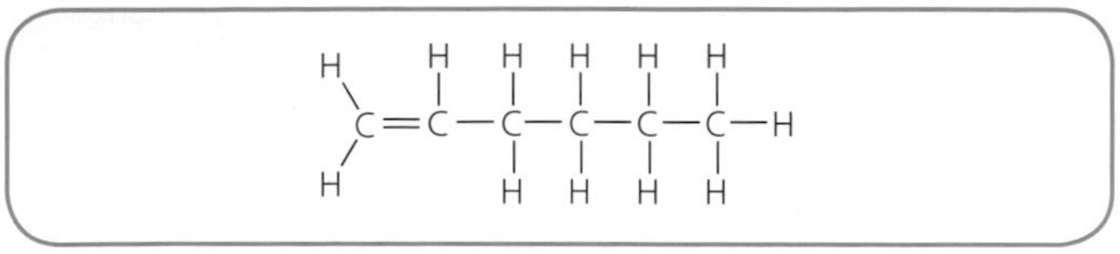

Figure 4 Hex-1-ene

Cyclohexane is an example of a cycloalkane and has the formula C_6H_{12}. Its structure is:

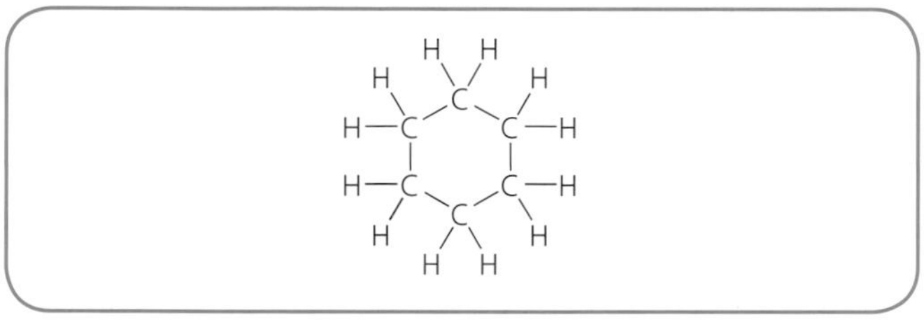

Figure 5 Cyclohexane

This information tells us that bottles A and B must contain hex-1-ene and cyclohexane. Both of these hydrocarbons fit the general formula C_nH_{2n} and, since they have the same molecular formula but a different structural formula, they are isomers.

Hexane must therefore be the hydrocarbon in bottle C since it does not have the formula C_6H_{12}.

As all the carbon-to-carbon bonds are single in cyclohexane and hexane, both are said to be saturated. As they are saturated, they would not be expected to take part in addition reactions. Hex-1-ene, however, has a carbon-to-carbon double bond so is said to be unsaturated and could take part in addition reactions.

Shaking a small sample of each of hydrocarbons A and B with bromine solution would allow the student to work out which one was hex-1-ene since only hex-1-ene would decolourise bromine solution. An equation for the reaction is shown in Figure 6.

This chemical test would allow you to label one of the bottles with formula C_6H_{12} as hex-1-ene and the other bottle with formula C_6H_{12} as cyclohexane.

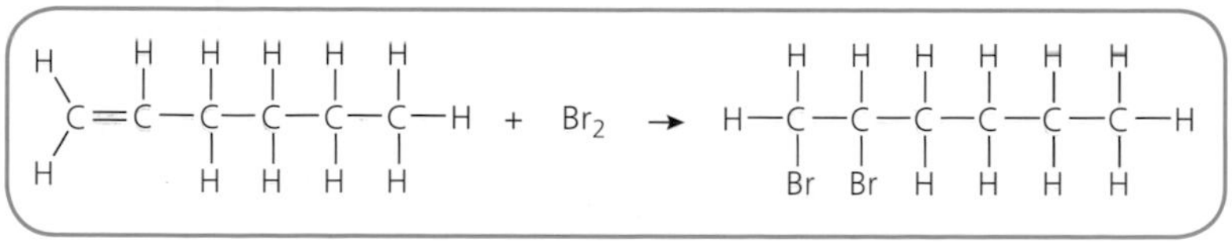

Figure 6 Reaction of bromine with hex-1-ene

Three questions to try for yourself

1* A group of students were given strips of aluminium, iron, tin and zinc.

   Using your knowledge of chemistry, suggest how the students could identify each of the four metals.

2* Nitrogen, phosphorus and potassium are elements essential for plant growth. A student was asked to prepare a dry sample of a compound which contained **two** of these elements. The student was given access to laboratory equipment and the chemicals in the table.

   Using your knowledge of chemistry, comment on how the student could prepare their dry sample.

Chemical	Formula
ammonium hydroxide	NH_4OH
magnesium nitrate	$Mg(NO_3)_2$
nitric acid	HNO_3
phosphoric acid	H_3PO_4
potassium carbonate	K_2CO_3
potassium hydroxide	KOH
sodium hydroxide	$NaOH$
sulfuric acid	H_2SO_4
water	H_2O

3 A student saw a silver brooch, which was thought to come from the time of the Roman Empire about 2000 years ago, in the local museum. The student had been studying radioisotopes, and learned that carbon-14 could be used to establish the age of objects such as wood, natural fibres and even human remains. The student wondered whether the isotope silver-108 (which has a half-life of 418 years) could be used to date the brooch more accurately.

   Using your knowledge of chemistry, explain whether silver-108 could be used to find the age of the brooch.

Glossary

A

Acid: A substance that produces hydrogen ions (H^+) when dissolved in water.

Addition polymerisation: The reaction between many small, unsaturated molecules (monomers) to form one large molecule (a polymer) and nothing else.

Addition reaction: A chemical reaction where molecules add across a double bond.

Alcohol: A compound containing an –OH group.

Alkali: A soluble base that produces hydroxide ions (OH^-) when dissolved in water.

Alkanes: A family of hydrocarbons where all the members are saturated and have the general formula C_nH_{2n+2}.

Alkenes: A family of hydrocarbons where all the members are unsaturated, they contain one carbon-to-carbon double bond and have the general formula C_nH_{2n}.

Alpha (α) particle: A charged particle consisting of two protons and two neutrons emitted by some radioisotopes. It is identical to a helium nucleus and is represented as $_2^4He$.

Atom: The smallest quantity of an element that can take part in a chemical reaction.

Atomic lattice: A large three-dimensional arrangement of atoms held together by covalent bonds.

Atomic number: The number of protons in the nucleus of an atom of an element.

B

Base: A substance that will neutralise an acid.

Battery: A series of chemical cells joined together (often the words cell and battery are used interchangeably).

Beta (β) particle: Charged particle consisting of a single electron emitted from the nucleus of some radioisotopes. Represented as $_{-1}^0e$.

Burette: A graduated piece of apparatus used to measure accurately the volume of liquid or solution dispensed during an experiment.

C

Calorimeter: Equipment used to measure the energy released in a chemical reaction.

Carboxylic acid: An acidic compound that contains the carboxyl group (–COOH).

Catalyst: A substance that will speed up a chemical reaction but is not used up during the process.

Ceramic: A compound of a metal and a non-metal that has gained certain properties by being heated and then cooled.

Combustion: The process of burning, in which a substance reacts with oxygen.

Compound: A substance made up of two or more elements chemically joined.

Concentration: A measure of how much solute is dissolved in a solvent, measured in $mol\,l^{-1}$.

Concordant: In agreement, consistent. In a titration, two results within $0.2\,cm^3$ of each other.

Covalent bond: A shared pair of electrons between two non-metal atoms.

Covalent network: A giant network of non-metal atoms held together by covalent bonds.

Cycloalkanes: A family of hydrocarbons where all the members are saturated and have the general formula C_nH_{2n}.

D

Decompose: To break down or cause to break down a compound into its component elements or simpler substances.

Deionised water: Water which has had all unwanted solute ions removed from it. A few H^+ and OH^- ions are present due to the partial dissociation of water.

Dilution: Addition of more solvent to a solution to decrease its concentration.

Dissociate: To break up or separate.

E

Electrochemical series (ECS): A list of metals (and hydrogen) in order of their ability to lose electrons and form ions in solution.

Electrolysis: The process which occurs when a direct current of electricity is passed through a molten electrolyte or an electrolyte solution which results in decomposition.

Electrolyte: A compound that conducts owing to the movement of ions either when dissolved in water or melted.

Electron: A negatively charged particle with a relative mass of zero, which orbits the nucleus of an atom.

Element: A substance that consists of atoms with the same number of protons in their nuclei.

Endothermic reaction: A reaction in which energy is taken in from the surroundings.

Enzyme: A biological catalyst.

Excess: A quantity of a reactant which is more than that required in a reaction.

Exothermic reaction: A reaction in which energy is released to the surroundings.

F

Feedstock: A substance obtained from a raw material which is then used to manufacture another substance.

Fermentation: The reaction which changes sugar into ethanol.

Functional group: A group of atoms responsible for the characteristic reactions of a particular compound. For example, the hydroxyl group (OH) in alcohols and the carboxyl group (COOH) in carboxylic acids.

G

Gamma (γ) rays: High-frequency, and high-energy, electromagnetic radiation emitted by radioactive substances.

Gram formula mass (GFM): The mass, in grams, of one mole of a substance.

Group ion: A charged particle that contains more than one type of atom (such as permanganate, MnO_4^-).

H

Half-life: The time in which the activity of a radioisotope decays by half, or in which half of its atoms disintegrate.

Homologous series: A family of compounds with the same general formula and similar chemical properties that show a gradual change in physical properties. The alkanes are an example of a homologous series.

Hydration: A chemical reaction in which a molecule of water is added to an unsaturated hydrocarbon producing an alcohol.

Hydrogenation: An addition reaction where hydrogen is added across a double bond.

I

Indicator: A substance whose colour changes depending on pH.

Ion: A charged particle.

Ion–electron half equation: Equation that shows either the loss of electrons (oxidation) or the gain of electrons (reduction).

Ionic bond: The electrostatic force of attraction between positive ions and negative ions.

Ionic lattice: A large arrangement of ions held together by ionic bonds.

Isomers: Compounds with the same molecular formula but different structural formulae.

Isotopes: Atoms with the same atomic number but different mass numbers.

M

Macroscopic: Large enough to be seen or examined by the unaided eye.

Malleable: A physical property of metals. A malleable material can be shaped by hammering or rolling.

Mass number: Equal to the number of protons plus neutrons in an atom or ion.

Meniscus: The visible curve on the upper surface of a liquid that is caused by surface tension when the liquid is held in a container.

Metallic bonding: Each atom in the metal element loses its outer electrons to form positive ions. These ions pack together in a regular crystalline arrangement and the free (delocalised) electrons move freely through the structure. The electrostatic attraction of the positive ions for the delocalised electrons binds the ions together and is called metallic bonding.

Mole: The relative atomic mass of an element expressed in grams (twice the RAM for diatomic elements); the sum of the RAMs of the elements in a compound (taking account of the number of atoms or ions of each element its formula contains), expressed in grams.

Molecule: Two or more atoms held together by covalent bonds.

Molten: Describes a liquid formed by melting a solid.

Monomers: Relatively small molecules that can join together to produce a very large molecule (a polymer) by a process called polymerisation.

N

Neutralisation: A reaction in which the pH moves towards 7.

Neutron: A neutral particle with a relative mass of 1 found in the nucleus of an atom.

Nuclear fission: Splitting nuclei by bombarding them with slow-moving neutrons.

Nuclear fusion: Light nuclei combining to form heavier nuclei and producing a lot of energy.

Nucleus: The positively charged centre of an atom which contains the neutrons and protons.

O

Oxidation: A reaction in which electrons are lost or oxygen is gained.

P

pH: A measure of how acidic or alkaline a substance is.

Phenyl group: A group of carbon and hydrogen atoms, formula C_6H_5-.

Pickling: A method used to keep food from spoiling. It involves storing the food in vinegar, which contains ethanoic acid, to stop bacteria and fungi growing.

Piezoelectric: Crystals that acquire an electrical charge when twisted, distorted, compressed or put under any mechanical stress are said to be piezoelectric.

Pipette: A piece of apparatus used to measure a specific volume of liquid or solution.

Polymer: A very large molecule formed by the joining together of many smaller molecules (monomers).

Precipitation: A reaction in which two solutions react to produce an insoluble product.

Proton: A positively charged particle with a relative mass of 1 found in the nucleus of an atom.

R

Radioisotopes: Atoms that emit radiation because their nuclei are unstable. Three types of radiation, α, β or γ, may be emitted.

Rate: A measure of how quickly a chemical reaction is progressing.

Redox reaction: A reaction that involves both reduction and oxidation.

Reducing agent: A substance that gives electrons away to another substance or that removes oxygen from it.

Reduction: A process in which electrons are gained or oxygen is lost.

Relative atomic mass (RAM): The average mass of the isotopes of an element.

Reversible: A reaction which occurs in a forward direction (reactants forming products) and simultaneously in a backward direction (where products react to form the original reactants).

S

Saturated: A compound that cannot have any more atoms added to it. All the bonds are single.

Solute: The substance that is dissolved in a solution.

Solution: The product of dissolving a solute in a solvent.

Solvent: The substance in which a solute is dissolved to make a solution.

Spectator ion: An ion that is present as both a reactant and a product in a chemical equation. Spectator ions take no part in the reaction.

Standard flask: A piece of laboratory glassware calibrated to contain a precise volume at a particular temperature.

Standard solution: A solution which contains a precisely known quantity of solute, used to determine the concentration of another solution by titration.

Strong acid: An acid which dissociates completely into ions in aqueous solution.

Sucrose: A sugar which has the formula $C_{12}H_{22}O_{11}$ found in many plants such as sugar cane.

Systematic naming: The process by which a compound is given a chemical name (systematic name), obtained by following a set of prescribed rules.

T

Tetrahedral: A tetrahedral molecule has a central atom bonded to four other atoms that are positioned at the corners of a regular tetrahedron.

Thermoplastic: A plastic that softens on heating and can be reshaped.

Titration: An experiment that will accurately measure the volume of liquid or solution required for a particular reaction.

Trigonal pyramid: A 3-dimensional molecular shape with one atom at the apex connected to three other atoms of a triangular base.

U

Unsaturated: A compound that can have more atoms added to it. It will contain at least one carbon-to-carbon double bond.

V

Valency: The number of bonds that an atom or ion can form.

Variable: A factor that can be changed in a chemical reaction that will affect the rate.

W

Weak acid: An acid which only partially dissociates into ions in aqueous solution.

Index

Answers

1 Rates of reaction

End-of-chapter questions (pages 8–10)

1 A

2 A

3 B

4 A

5 B

6 A powdered catalyst will have a greater surface area due to the smaller particle size making it more efficient.

7 **a)**

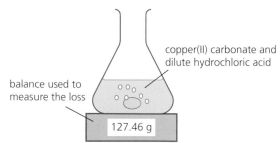

balance used to measure the loss

copper(II) carbonate and dilute hydrochloric acid

127.46 g

b)

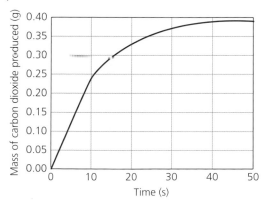

c) The line that is plotted becomes horizontal, indicating that even though time is progressing, there is no increase in the mass of gas that is produced.

d) $0.024\,g\,s^{-1}$

8 **a)** Hydrogen gas will burn with a squeaky pop.

b) Any two from the following three: use a lower concentration of hydrochloric acid; use larger particles of zinc; lower the temperature of the reaction mixture.

c)

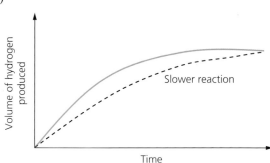

Slower reaction

9 **a)** $0.0165/0.017\,mol\,l^{-1}\,min^{-1}$

b) $0.0038/0.004\,mol\,l^{-1}\,min^{-1}$

10 **a)**

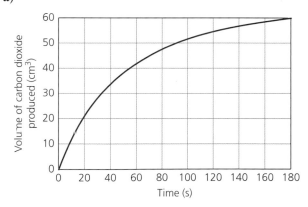

b) $55\,cm^3$

c)

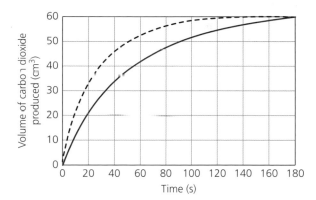

d)

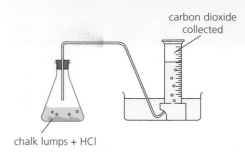

carbon dioxide
collected

chalk lumps + HCl

2 Atomic structure

End-of-chapter questions (page 16)

1 A
2 B
3 A
4 B
5 D
6 **a)**

Particles	Number
protons	47
electrons	46
neutrons	60

b) Isotopes
c) Equal quantities of each isotope.

7 **a)**

Isotope	Number of protons	Number of electrons	Number of neutrons
$^{35}_{17}Cl$	17	17	18
$^{37}_{17}Cl$	17	17	20

b) The isotope with a mass of 35 is the more abundant.

8

Isotope	Mass number	Atomic number	Number of protons	Number of electrons	Number of neutrons
$^{31}_{15}P$	31	15	15	15	16
$^{39}_{19}K$	39	19	19	19	20
$^{15}_{7}N$	15	7	7	7	8

9 **a)** $^{41}_{20}Ca$
   b) Protons = 20; neutrons = 21
   c) The proton to electron ratio is 1:1 **or** there are equal numbers of protons and electrons

10

Ion	Number of protons	Number of electrons	Number of neutrons
$^{35}_{17}Cl^{-}$	17	18	18
$^{16}_{8}O^{2-}$	8	10	8
$^{40}_{20}Ca^{2+}$	20	18	20

3 Bonding, structure and properties

End-of-chapter questions (page 23)

1 D
2 B
3 B
4 C
5 **a)** A covalent bond is a shared pair of electrons between two non-metals. The atoms are held together because of the electrostatic force of attraction between the positive nuclei of each atom and the negatively charged electrons.

b)

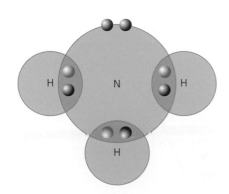

c)

methane

ammonia

water

6

Substance	Bonding and structure
C	ionic
B	covalent network
A	covalent molecular

7 Silicon dioxide has a covalent network structure. This means that a lot of energy is required to break the strong covalent bonds that hold the atoms together, meaning that very high temperatures are required to turn silicon dioxide into a gas. Carbon dioxide has a covalent molecular structure. This means that only the weak forces of attraction between the molecules are broken when it is turned into a gas. This is why it is a gas at room temperature.

8 Salt is an ionic compound. In solution, the ions are free to move. Glucose is a covalent molecular compound. Since molecules are not charged particles, a current cannot flow.

4 Formulae and reacting quantities

In-text questions

1 a) Magnesium and oxygen
b) Beryllium and bromine
c) Potassium, manganese and oxygen
d) Lead, nitrogen and oxygen
e) Sodium, hydrogen, carbon and oxygen

2 a) One atom of hydrogen, one atom of fluorine
b) One atom of sulfur, three atoms of oxygen
c) Two atoms of nitrogen, four atoms of oxygen
d) One atom of phosphorus, three atoms of chlorine
e) Two atoms of hydrogen, one atom of oxygen

3 a) CO
b) BBr_3
c) PCl_5
d) N_2O_3
e) XeF_6

4 a) Li_2S
b) KCl
c) Ca_3P_2
d) BBr_3
e) Mg_3N_2

5 a) $Mg(OH)_2$
b) $NaHCO_3$
c) $KMnO_4$
d) $Al_2(SO_4)_3$
e) $(NH_4)_3PO_4$

6 a) $Mg^{2+}(OH^-)_2$
b) $Na^+HCO_3^-$
c) $K^+MnO_4^-$
d) $(Al^{3+})_2(SO_4^{2-})_3$
e) $(NH_4^+)_3PO_4^{3-}$

7 a) Au_2O_3
b) $SnCl_4$
c) Fe_2S_3
d) $CuBr$
e) PbI_2

8 a) $C_2H_4(g) + 3O_2(g) \longrightarrow 2CO_2(g) + 2H_2O(l)$
b) $3Ag_2S(aq) + 2Al(s) \longrightarrow Al_2S_3(aq) + 6Ag(s)$
c) $C_6H_{12}O_6(s) + 6O_2(g) \longrightarrow 6CO_2(g) + 6H_2O(l)$
d) $2H_3PO_4(aq) + 3Mg(OH)_2(s) \longrightarrow$
$$Mg_3(PO_4)_2(s) + 6H_2O(l)$$
e) $2CO(g) + O_2(g) \longrightarrow 2CO_2(g)$

9 a) 64 g
b) 16 g
c) 158 g
d) 100 g
e) 148.5 g

10 a) 186 g
b) 84 g
c) 11.5 g
d) 245 g
e) 477.5 g

11 a) 3 moles
b) 1.5 moles
c) 0.1 moles
d) 2.5 moles
e) 4 moles

12 a) 0.25 g
b) 1.4 g
c) 2.2 g
d) 90 g

13 a) $0.5\,mol\,l^{-1}$
b) $1.5\,mol\,l^{-1}$
c) $1\,mol\,l^{-1}$
d) $2\,mol\,l^{-1}$
e) $2.4\,mol\,l^{-1}$

14 a) $0.2\,mol\,l^{-1}$
b) $1\,mol\,l^{-1}$
c) $0.4\,mol\,l^{-1}$
d) $2\,mol\,l^{-1}$
e) $0.5\,mol\,l^{-1}$

15 a) 75%
   b) 40%
   c) 48%
   d) 32.4%
   e) 17.2%

5 Acids and bases

End-of-chapter questions (pages 45–46)

1 C
2 D
3 C
4 C
5 a) nitric acid + potassium oxide $\longrightarrow$ **potassium nitrate + water**
  b) sulfuric acid + lithium carbonate $\longrightarrow$ **lithium sulfate + water + carbon dioxide**
  c) sodium hydroxide + nitric acid $\longrightarrow$ **sodium nitrate + water**
  d) hydrochloric acid + ammonium hydroxide $\longrightarrow$ **ammonium chloride + water**
6 a) Pipette
  b) Neutralisation
  c) 0.0025 moles
  d) 24.8 cm^3
7 a) potassium carbonate + sulfuric acid $\longrightarrow$ potassium sulfate + water + carbon dioxide
  b) $K_2CO_3 + H_2SO_4 \longrightarrow K_2SO_4 + H_2O + CO_2$
  c) 2 mol l^{-1}
8 a) A base is a substance that will neutralise an acid.
  b) An alkali is a soluble base.
9 0.12 mol l^{-1}
10 0.025/0.03 mol l^{-1}
11 0.143/0.14 mol l^{-1}
12 60 cm^3
13 16 cm^3
14 0.25 mol l^{-1}
15 0.0417/0.04 mol l^{-1}
16 300 cm^3
17 300 cm^3
18 3

6 Homologous series

In-text questions

1 a)

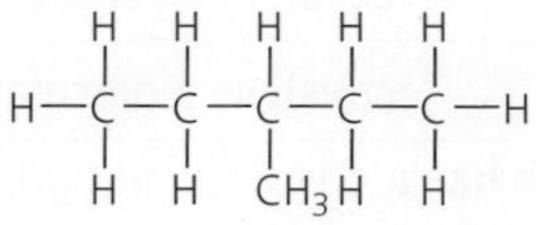

   b)

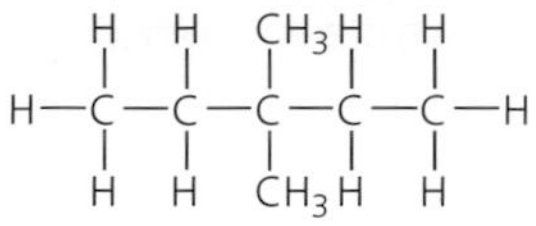

   c)

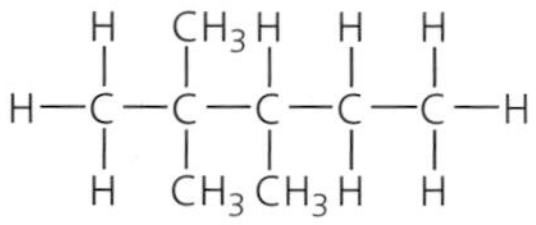

End-of-chapter questions (pages 59–60)

1 a) B and C
  b) B and E
  c) E
  d) A, D and F
  e) D
2 a) A and E
  b) A
  c) F
  d) C
  e) D and F
3 a)

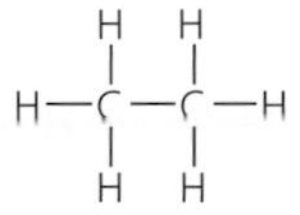

  b)

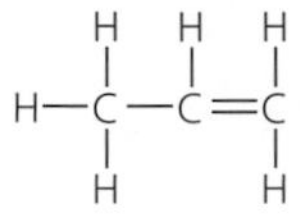

  c)

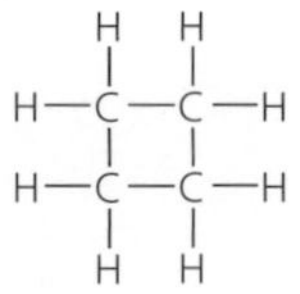

4 A

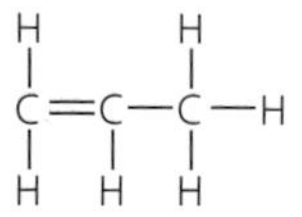

B

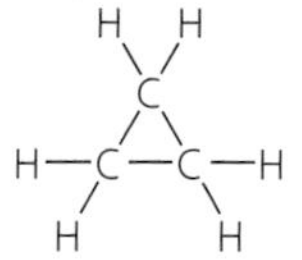

5 a)

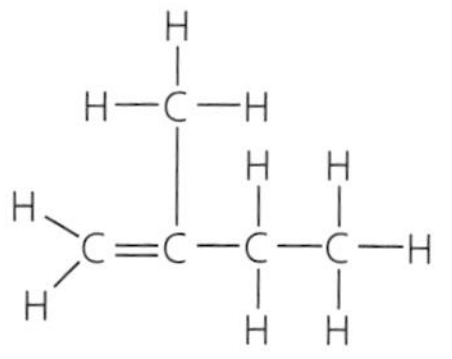

b)

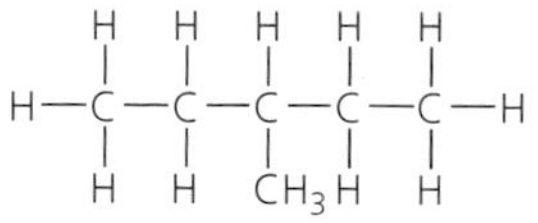

6 a) 2-methylhexane

b) 4,4-dimethylpent-2-ene

c) 2,3-dimethylbut-1-ene

7 a) C_nH_{2n+2}

b) i) $C_{12}H_{26}$

ii) $C_{39}H_{80}$

8 a)

Name	Structure	Molecular formula
ethyne	$H-C\equiv C-H$	C_2H_2
propyne	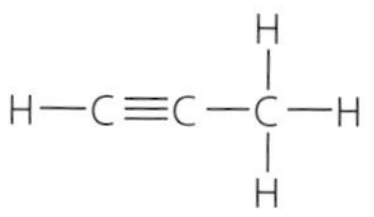	C_3H_4
but-1-yne	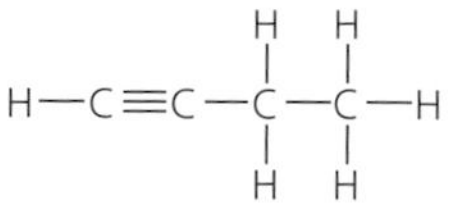	C_4H_6

b)

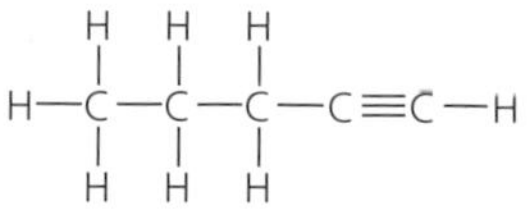

c) C_nH_{2n-2}

9 a) i) $C_nH_{2n}O$

ii) If the OH is on the end/first carbon atom, an aldehyde is obtained. If the OH is on the second carbon atom, a ketone is obtained.

b)

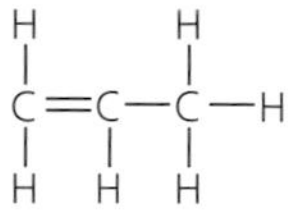

c)

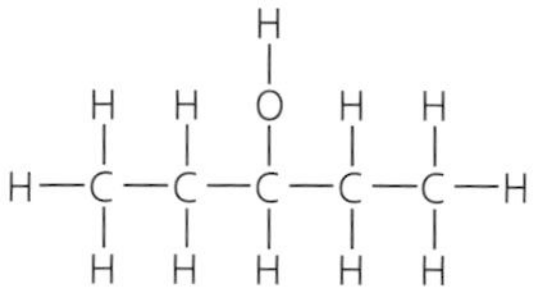

10 a) A family of compounds with the same general formula and similar chemical properties that show a gradual change in physical properties.

b) C_nH_{2n-2}

c) $C_5H_8Br_4$

d)

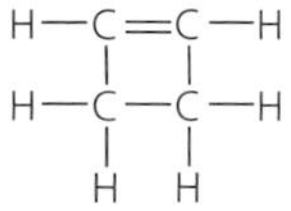

11 a)

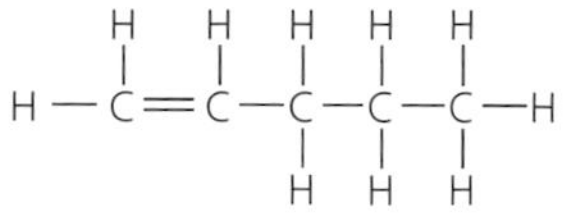

b)

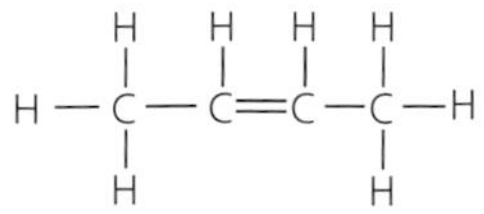

7 Everyday consumer products

In-text questions

1 a)

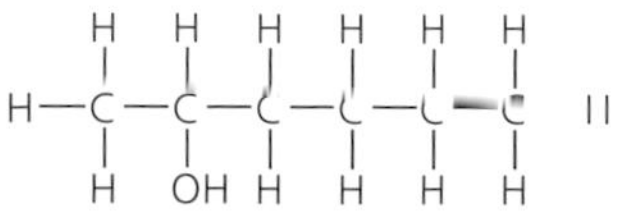

b)

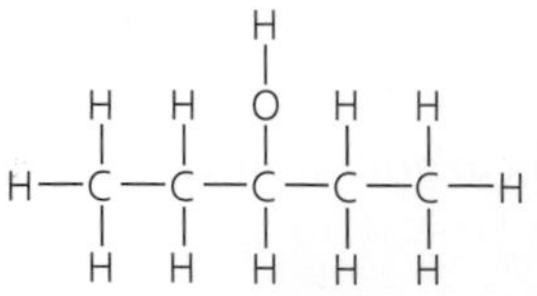

2 Hexan-4-ol would be incorrect because alcohols are named so that the –OH is attached to the lowest numbered carbon atom. Hexan-4-ol should be called hexan-3-ol.

3 **a)** Pentan-2-ol
   b) Heptan-3-ol

4 **a)**

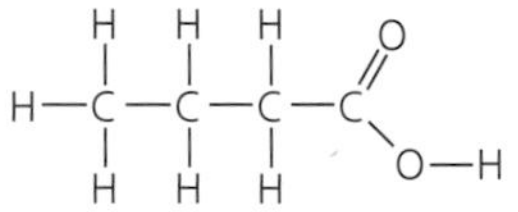

   b)

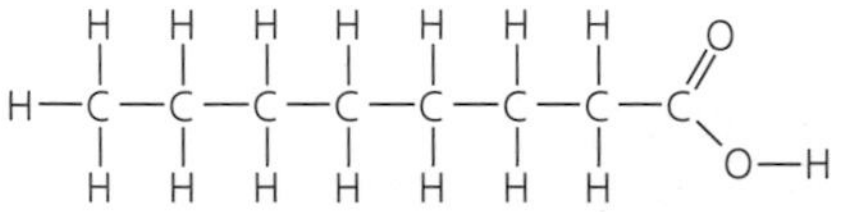

5 **a)** Pentanoic acid
   b) Heptanoic acid

6 $C_nH_{2n+1}COOH$ or $C_nH_{2n}O_2$

End-of-chapter questions (pages 69–70)

1 C
2 **a)** D
   b) C
   c) D
   d) A
   e) D
3 B
4

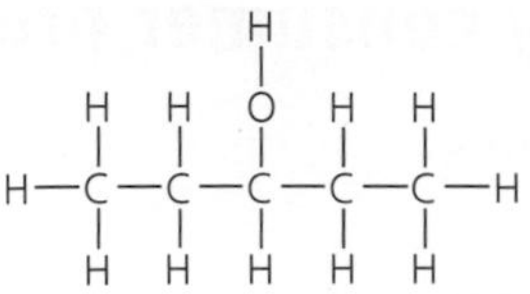

5

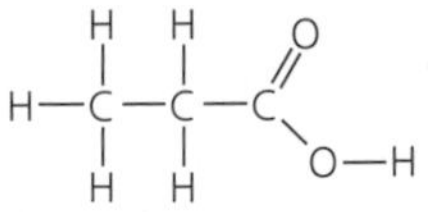

6 Can be used as cleaning chemicals. Can be used as food preservatives.

7 $C_2H_5OH + 3O_2 \longrightarrow 2CO_2 + 3H_2O$

8 Propanoic acid

9 **a)**

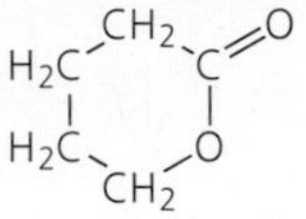

   b)

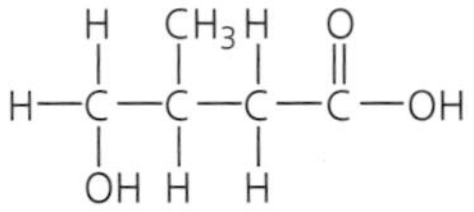

10 **a)** Add bromine solution to a sample of the lavender oil and shake. If it contains unsaturated compounds, the bromine solution will decolourise.
   b) –OH
   c)

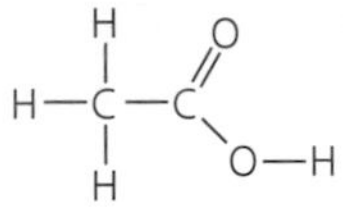

11 **a)** Potassium methanoate
   b) Lithium propanoate
   c) Sodium ethanoate
   d) Sodium butanoate
   e) Hexanoic acid
   f) Butanoic acid
   g) Methanoic acid
   h) Ethanoic acid

8 Energy from fuels

In-text questions

1 **a)** $CH_4 + 2O_2 \longrightarrow CO_2 + 2H_2O$
   b) $C_4H_{10} + 6\frac{1}{2}O_2 \longrightarrow 4CO_2 + 5H_2O$
   c) $C_6H_{14} + 9\frac{1}{2}O_2 \longrightarrow 6CO_2 + 7H_2O$
2 **a)** $C_2H_5OH + 3O_2 \longrightarrow 2CO_2 + 3H_2O$
   b) $C_3H_7OH + 4\frac{1}{2}O_2 \longrightarrow 3CO_2 + 4H_2O$
   c) $C_6H_{13}OH + 9O_2 \longrightarrow 6CO_2 + 7H_2O$
3 72 g
4 176 g
5 4.5 g
6 44 g
7 76.67/76.7 g

8 64 g
9 14.28/14.3 g
10 10 g

End-of-chapter questions (pages 78–79)

1 a) Exothermic; 60 kJ released
   b) Endothermic; 70 kJ taken in
2 180 g
3 1760 g
4 98.29/98.3 g
5 13.75 g
6 47.92/47.9 g
7 297.2 kJ
8 1933.25/1933.3 kJ
9 a) Measure the temperature of the water at the start of the experiment and measure the highest temperature of the water. Measure the mass of the spirit burner before burning and after burning. Measure the volume/mass of water.
   b) 208.65/208.6 kJ
   c) Less heat is lost to the surroundings and the combustion is more complete.
10 145.45 kg

9 Metals

In-text questions

1 a) The electrostatic attraction of positive metal ions for delocalised electrons present in the lattice.
   b) Malleable means something that can be shaped by hammering or rolling.
2 Agree. Covalent bonding involves sharing of electrons between bonded atoms and all the metal ions in the lattice share their outer (delocalised) electrons. Ionic bonding involves loss and gain of electrons and metal atoms in the lattice lose their outer electrons, which become delocalised.
3 a) Mercury, Hg
   b) Lattice means a regular three-dimensional arrangement of particles.
   c) The metal ions in the lattice change places and slide past each other to adapt to a new shape. This happens because all the ions in a metal lattice are the same size and have the same charge.
4 a) $4K + O_2 \longrightarrow 2K_2O$
   b) $4K + O_2 \longrightarrow 2(K^+)_2O^{2-}$

5 a) $Ca + 2H_2O \longrightarrow Ca(OH)_2 + H_2$
   b) $Ca(s) + 2H_2O(l) \longrightarrow$
      $Ca^{2+}(aq) + 2OH^-(aq) + H_2(g)$
6 a) $Mg + 2HCl \longrightarrow MgCl_2 + H_2$
   b) $Mg(s) + 2H^+(aq) + 2Cl^-(aq) \longrightarrow$
      $Mg^{2+}(aq) + 2Cl^-(aq) + H_2(g)$
7 a) C A B
   b) C = K, Na, Li, Ca or Mg; A = Al, Zn, Fe, Sn or Pb; B = Cu, Hg, Ag or Au
8 a) Group 3
   b) GaN
   c) Silicon
9 Any two elements, with correct symbols, from the Lanthanide series.
10 a) $2HgO \longrightarrow 2Hg + O_2$
    b) $2Hg^{2+}O^{2-}(s) \longrightarrow 2Hg(l) + O_2(g)$
11 a) $2CuO + C \longrightarrow 2Cu + CO_2$
    b) $2Cu^{2+}O^{2-}(s) + C(s) \longrightarrow 2Cu(s) + CO_2(g)$
    c) Carbon
12 a) Oxidation
    b) Reduction
    c) Reduction
13 B
14 a) $Fe(s) \longrightarrow Fe^{2+}(aq) + 2e^-$ [O]
       $2H_2O(l) + O_2(g) + 4e^- \longrightarrow 4OH^-(aq)$ [R]
    b) $2Fe(s) + 2H_2O(l) + O_2(g) \longrightarrow$
       $2Fe^{2+}(aq) + 4OH^-(aq)$
15 a) From Mg to Fe through the wires and meter
    b) From Sn to Ag
    c) From Zn to Ni
    d) From Al to Pb
16 a) Fe/Cu
    b) Fe/Ag
    c) Mg/Au
    d) Al/Hg

End-of-chapter questions (pages 94–95)

1 a) i) $2CuO + C \longrightarrow 2Cu + CO_2$
      ii) $CuO + CO \longrightarrow Cu + CO_2$
   b) Magnesium is too reactive.
   c) Mercury, silver or gold
2 a) Measure out some dilute sulfuric acid into a beaker. Add small quantities of zinc until fizzing can no longer be seen. Filter off the excess zinc. Evaporate the water from the filtrate.
   b) $Zn + H_2SO_4 \longrightarrow ZnSO_4 + H_2$
   c) $Zn(s) + 2H^+(aq) + SO_4^{2-}(aq) \longrightarrow$
      $Zn^{2+}(aq) + SO_4^{2-}(aq) + H_2(g)$

3 a) Electrons flow from X to Y.

b) Reduction

c) $2Fe^{3+}(aq) + SO_3^{2-}(aq) + H_2O(l) \longrightarrow$
$2Fe^{2+}(aq) + SO_4^{2-}(aq) + 2H^+(aq)$

4 a) 3+

b) $Al^{3+} + 3e^- \longrightarrow Al$

5 B

6 a) Electrode A/positive electrode

b) $2Na^+ + 2H^- \longrightarrow 2Na + H_2$

c) Metal elements contain delocalised electrons that carry the current.

7 A

8 a) $Zn(s) + Cu^{2+}(aq) \longrightarrow Zn^{2+}(aq) + Cu(s)$

b) i) $Fe(s) \longrightarrow Fe^{2+}(aq) + 2e^-$

ii) Oxidation

10 Plastics

End-of-chapter questions (pages 101–103)

1 Buta-1,3-diene

2 a) The linking together of many small unsaturated molecules to form one large saturated molecule.

b)

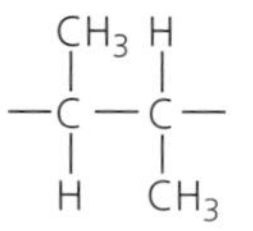

c)

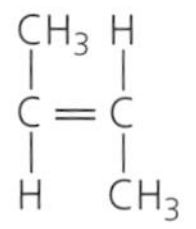

d) But-2-ene

3 a)

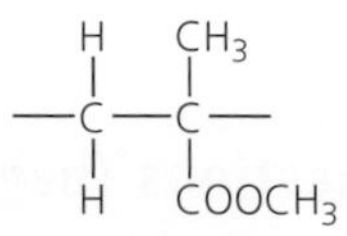

b)

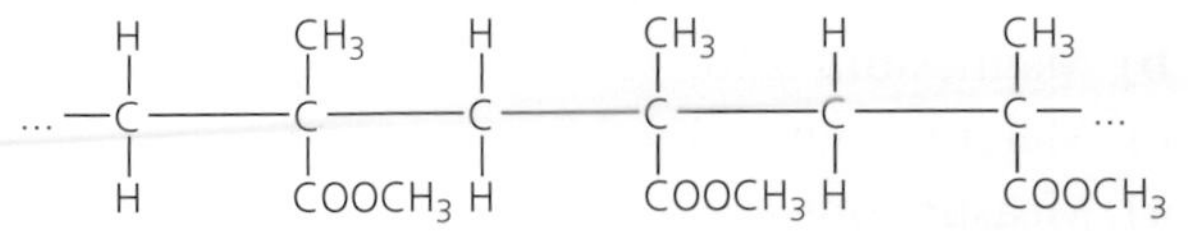

4 a) Propene and poly(propene)

b) Alkenes

5

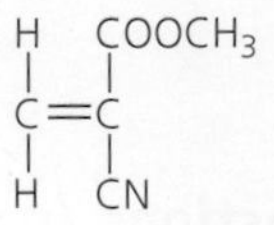

6 a) An addition polymer is a very big molecule formed when many small, unsaturated molecules (monomers) combine to form the polymer and nothing else.

b)

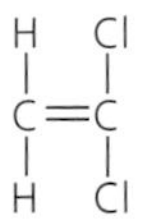

7 a)

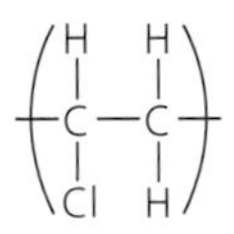

b)

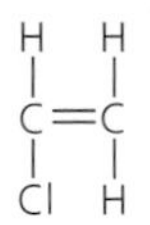

8

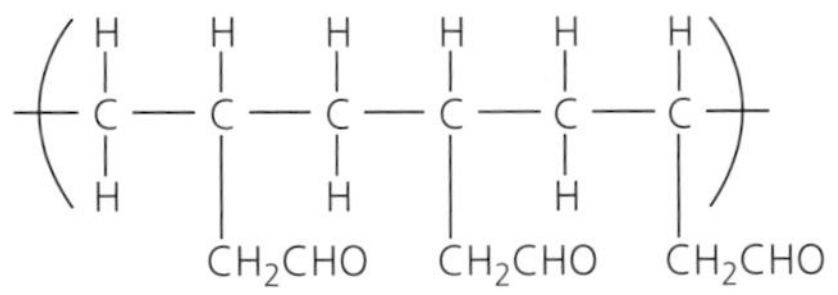

9

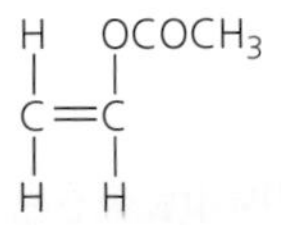

10 a) An addition polymer

b)

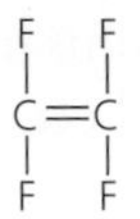

11 a)

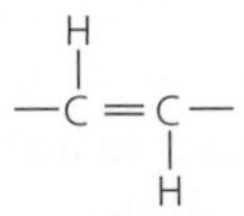

b)

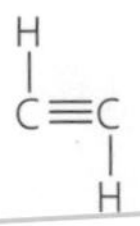

11 Fertilisers

In-text questions

1 $Ca(OH)_2 + (NH_4)_2SO_4 \longrightarrow CaSO_4 + 2NH_3 + 2H_2O$
2 Dimerise means two monomer units combining together.
3 N_2O_4
4 Dinitrogen monoxide
5 $NO_2 + O_2 \longrightarrow NO + O_3$; O_3 is called ozone; it forms in the upper atmosphere and absorbs some of the harmful ultra-violet radiation emitted by the Sun.
6 To increase the surface area and make the catalyst more efficient.
7 $4NH_3(g) + 5O_2(g) \longrightarrow 4NO(g) + 6H_2O(g)$
$2NO(g) + O_2(g) \longrightarrow 2NO_2(g)$
$4NO_2(g) + O_2(g) + 2H_2O(l) \longrightarrow 4HNO_3(aq)$
8

Acid	Formula of acid	Ions present in aqueous solution
phosphoric	H_3PO_4	$3H^+(aq)$ and $PO_4^{3-}(aq)$
sulfuric	H_2SO_4	$2H^+(aq)$ and $SO_4^{2-}(aq)$
nitric	HNO_3	$H^+(aq)$ and $NO_3^-(aq)$
hydrochloric	HCl	$H^+(aq)$ and $Cl^-(aq)$

9 $Ca_3(PO_4)_2 + 3H_2SO_4 \longrightarrow 2H_3PO_4 + 3CaSO_4$
10 a) $HCl + KOH \longrightarrow KCl + H_2O$
   b) Chloride ions and potassium ions
   c) $H^+ + OH^- \longrightarrow H_2O$
11 a) $KOH + HNO_3 \longrightarrow KNO_3 + H_2O$
   b) $NH_3 + HCl \longrightarrow NH_4Cl$
   c) $3NH_3 + H_3PO_4 \longrightarrow (NH_4)_3PO_4$
12 a) potassium hydroxide + sulfuric acid $\longrightarrow$
                              potassium sulfate + water
   $2KOH + H_2SO_4 \longrightarrow K_2SO_4 + 2H_2O$
   b) sodium hydroxide + nitric acid $\longrightarrow$
                              sodium nitrate + water
   $NaOH + HNO_3 \longrightarrow NaNO_3 + H_2O$
   c) potassium hydroxide + phosphoric acid $\longrightarrow$
                              potassium phosphate + water
   $3KOH + H_3PO_4 \longrightarrow K_3PO_4 + 3H_2O$
   d) ammonia solution + sulfuric acid $\longrightarrow$
                              ammonium sulfate + water
   $2NH_4OH + H_2SO_4 \longrightarrow (NH_4)_2SO_4 + 2H_2O$

End-of-chapter questions (pages 112–113)

1 a) 16.47/16.5%
   b) 17.07/17.1%
   c) 21.21/21.2%
   d) 35%
2 52.35/52.3%
3 a) i) 60
      ii) 46.67/46.7%
   b) $2NH_3 + CO_2 \longrightarrow CO(NH_2)_2 + H_2O$
4 a) Ostwald process
   b) Platinum
   c) KNO_3
5 a) i) Hydrogen
      ii) Haber process
   b) At temperatures below 450 °C the reaction is too slow to be cost effective.
   c) Ammonium nitrate
   d) In countries with a high rainfall Nitram would be washed out of the soil before it had a chance to be absorbed by the crops.
6 a) Any alkali, e.g. soda lime, NaOH, KOH etc.
   b) i) Platinum
      ii) The reaction is exothermic so it provides its own heat.
7 Nitrogen: 21.21/21.2%
   Phosphorus: 23.48/23.5%
8 a) i) 20%
      ii) 26.49/26.5%
   b) Calcium phosphate is insoluble in water so could not be absorbed by the crops. If calcium dihydrogen phosphate is used as a fertiliser, it must be soluble in water; it also contains a higher percentage of phosphorus.

12 Nuclear chemistry

In-text questions

1 ^{39}K protons = 19, neutrons = 20; ^{238}U protons = 92, neutrons = 146
2 Uranium

3

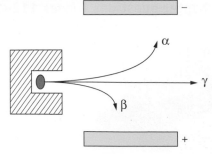

Alpha particles are positively charged so are attracted to the negative electrode. Beta particles are negatively charged so are attracted to the positive electrode. Gamma rays have no charge so are unaffected by the electric field.

4

Name	Alpha (α)	Beta (β)	Gamma (γ)
symbol	^4_2He	$^0_{-1}e$	γ
mass	4 mass units	0 mass units (approximately)	none
charge	positive (2+)	negative (1–)	none
penetrating power	low (sheet of paper)	higher (thin metal sheet)	very high (several cm of lead)

5 a) C
   b) A
   c) D
6 a) $a = 234$; $b = 90$, X is thorium/Th
   b) $a = 32$; $b = 16$, X is sulfur/S
   c) $a = 216$; $b = 84$, X is polonium/Po
7 Change 1 is alpha emission because the mass number has decreased by 4 units and the atomic number has decreased by 2 units.
   Change 2 is beta emission because the mass number is unchanged and the atomic number has increased by 1 unit.
8 a) $^3_1H \longrightarrow {}^3_2He + {}^0_{-1}e$
   b) $^{224}_{88}Ra + {}^4_2He \longrightarrow {}^{228}_{90}Th$
   c) $^{60}_{27}Co \longrightarrow {}^{60}_{27}Co + \gamma$
9 7.5 years
10 8 days
11 D

End-of-chapter questions (page 122)

1 a) $^{14}_6C \longrightarrow {}^{14}_7N + {}^0_{-1}e$
   b) Fossil fuels are too old.

2 a) $^{35}_{16}S \longrightarrow {}^{35}_{17}Cl + {}^0_{-1}e$
   b) 174 days
3 a) $a = 234$, $b = 91$, X is protactinium
   b) $a = 206$, $b = 82$, X is lead
   c) $a = 214$, $b = 83$, X is bismuth
4 a) Cobalt-60 or iodine-131
   b) Americium-241 has a long half-life and is an alpha emitter so radiation is absorbed by the smoke detector's plastic cover.
   c) γ radiation has a high penetrating power and would pass through the paper too easily.
5 In the upper atmosphere, nitrogen atoms capture neutrons to produce the radioisotope carbon-14. The concentration of carbon-14 atoms in the atmosphere is fairly constant because, as they form, they combine with oxygen and the CO_2 produced is absorbed by green plants on the planet's surface. In living plants, the carbon-14 is slowly but continuously decaying and it is simultaneously being replaced by CO_2 absorbed during photosynthesis. When a plant dies, it stops taking in carbon dioxide, but the carbon-14 already in the plant keeps on decaying. If we assume that the ratio of carbon-14 to carbon-12 in trees today is the same as it was in the past, then by measuring the radioactivity due to carbon-14 in a sample of old wood and in the same type of new wood, scientists can estimate the age of the old wood.
6 a) Cosmic rays/radon/granite rock
   b) In the stars

13 Chemical analysis

In-text questions

1 a) Nickel(II) carbonate
   b) Barium sulfate
2 a) i) Any very soluble lead(II) compound, e.g. lead(II) nitrate, and any very soluble carbonate, e.g. ammonium carbonate, lithium carbonate, potassium carbonate or sodium carbonate.
      ii) Any correct equation using the compounds named in a) i), e.g.
$$Pb(NO_3)_2(aq) + K_2CO_3(aq) \longrightarrow PbCO_3(s) + 2KNO_3(aq)$$
      iii) $Pb^{2+}(aq) + CO_3^{2-}(aq) \longrightarrow Pb^{2+}CO_3^{2-}(s)$
   b) Precipitation

3

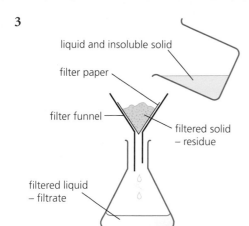

liquid and insoluble solid

filter paper

filter funnel

filtered solid
– residue

filtered liquid
– filtrate

4 a) The flasks get lighter because the reaction produces carbon dioxide gas which is lost to the atmosphere.

b) A measuring cylinder containing the volume of acid to be used in the reaction could be placed, next to the flask containing the marble chips, on the balance so that the total mass (of apparatus + reactants) could be measured before the reaction started; the empty measuring cylinder must be replaced on the balance after the reactants have been mixed.

5 a) The processed data in Table 13.5 was obtained by subtracting each mass recorded in Table 13.4 from the original total mass for that set of apparatus recorded at time zero.

b) A continuous variable is one which can be sub-divided into smaller units.

c) The lines become horizontal because both reactions have stopped.

d) i) $1.17\,g\,min^{-1}$

ii) $0.83\,g\,min^{-1}$

End-of-chapter questions (page 133)

1 A is copper(II) sulfate; B is potassium chloride; C is sulfuric acid; D is silver(I) nitrate

2 a) $Cu(OH)_2$

b) A fine white precipitate would form; the colour of the precipitate would be masked by the blue colour of the solution.

c) The sample to be tested is mixed with a few drops of hydrochloric acid. A flame test rod is dipped into the mixture then held in the edge of a blue Bunsen flame and the colour of the flame is recorded.

3 a) A pH meter

b) i) A pipette and a burette

ii)

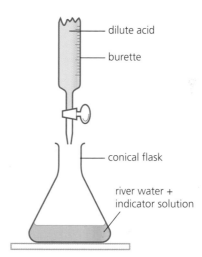

dilute acid

burette

conical flask

river water +
indicator solution

iii) When the river water had been neutralised, the indicator would change colour.

c) Accurate titre = $(14.5 + 14.6)/2 = 14.55\,cm^3$

d) River water is a mixture and there may be several substances in the mixture that react with the acid during the titration. It is not valid to assume that the only substance reacting with the acid is a substance that releases hydroxide ions.